Revolution

REVOLUTION

An Introduction to the

Art and Science of

Worldview Warfare

Donald Oliver Davis

Olive Informatics LLC

Huntsville Alabama

Revolution: an Introduction to the Art and Science of Worldview Warfare

Copyright © 2020 Donald Oliver Davis – Revised Edition, February 2021

Book Design, Photos, and Illustrations by Donald Oliver Davis

All rights reserved. No part of this book may be used or reproduced in any manner whatsoever without the written permission of the publisher, except in the case of brief quotations in a book review

ISBN: 978-1-7345454-0-1

Publisher's Cataloging-In-Publication Data
(Prepared by The Donohue Group, Inc.)

Names: Davis, Donald Oliver, 1982- author.

Title: Revolution : an introduction to the art and science of worldview warfare / Donald Oliver Davis.

Description: Huntsville, Alabama : Olive Informatics LLC, [2020] | Includes bibliographical references and index.

Identifiers: ISBN 9781734545401

Subjects: LCSH: Social engineering. | Military art and science--Social aspects. | Psychological warfare. | Spiritual warfare. | Information technology--Social aspects.

Classification: LCC HM668 .D38 2020 | DDC 301--dc23

Published by Olive Informatics LLC

Huntsville, AL

www.oliveinformatics.com

Old Testament Scripture references taken from the St. Athanasius Academy Septuagint™. Copyright © 2008 by St. Athanasius Academy of Orthodox Theology. Used by permission. All rights reserved.

New Testament Scripture references taken from the New King James Version®. Copyright © 1982 by Thomas Nelson, Inc. Used by permission. All rights reserved.

Scripture exegeted with the assistance of Gesenius' [1857] Hebrew-English Lexicon of the Old Testament and Thayer's [1996] Greek-English Lexicon of the New Testament – as well as the testimonies of countless other faithful brothers and sisters in the Church; past as well as present

To Mom, whose life taught me the profound wisdom in this:

"Love never fails. But whether there are prophecies, they will fail; whether there are tongues, they will cease; whether there is knowledge, it will vanish away. For we know in part and we prophesy in part. But when that which is perfect has come, then that which is in part will be done away. When I was a child, I spoke as a child, I understood as a child, I thought as a child; but when I became a man, I put away childish things. For now we see in a mirror, dimly, but then face to face. Now I know in part, but then I shall know just as I also am known. And now abide faith, hope, love, these three; but the greatest of these is love."

[1 Corinthians 13: 8-13]

CONTENTS

Introduction 1

Nature 9

    Technology I 14

    Universal Physis 16

    Deeper Levels 18

Music 29

    Technology II 32

    Universal Mousikē 34

    Deeper Levels 39

Spirituality 47

    Western Religion 50

    Greco-Roman Religion 55

    Phoenician Religion 57

    Babylonian Religion 59

    Egyptian Religion 61

Military Art 65

    The Art of Revolution 72

    The Art of Counter-Revolution 77

Conclusion 83

Bibliography 87

Index 131

Convento de las Dueñas and Catedral de Salamanca

Salamanca, Spain

INTRODUCTION

I spent the summer of 2005 searching for ways to foster understanding and peace among people of diverse religious, economic, political, and ethnic backgrounds – in Spain. Such was my mission as an Ambassadorial Scholar for district 6780 of Rotary International. During business hours, I studied Castilian Spanish at the ancient University of Salamanca. Outside of business hours, I eagerly immersed myself in Castilian society [Marcos, 2005]. Newfound (albeit, non-Castilian) friends helped me to open a bank account, obtain an apartment, and make daily visits to farmers' markets – just like a local. I was, at all times, learning important lessons about our world as well as myself.

One such lesson involved the profound ways in which distant pasts and near futures can coalesce. Indeed, Salamanca once hosted Francisco Franco's military headquarters – as well as Mussolini's and Hitler's military attachés – during the Spanish Civil War. Salamanca also hosted documents Franco's army stole from Catalans; documents later used to identify as well as oppress – even eradicate – Franco's Catalan opponents [Crameri, 2006]. Fast forward nearly a century, and some Castilians (identifying as descendants of said war's victors) are fighting to keep the documents in Salamanca. I found myself in the middle of this fight when I waded through crowds of impassioned Castilian protesters while walking to class one sunny day. I was surprised to sense the familiar sting of ethnic tension between people whose skin tones were indistinguishable, in my view. I was not surprised, over a decade later, to watch live television footage of violence during Catalonia's renewed fight for national [i] independence.

I later found myself in the middle of an intensifying global sociocultural issue: mass migration. During late-evening strolls on Salamanca's still lively, cobblestone streets, I increasingly noticed boisterous groups of young Spanish men uttering the word 'moro' as I passed. Although I did not initially assume they directed the word at me, I did sense a negative connotation. Curious about its meaning, I asked new friends to clarify. One friend was a Roman Catholic, Kenyan scholar who later married a Basque woman. He described it as a slang slur signifying many Spaniards' resentment against both the refugees of African ancestry now pouring into their country and the invaders of African ancestry who had once colonized and enslaved their ancestors. Since Roman times, these indigenous peoples of Northwestern Africa were referred to as 'Moors'.

Contention around similar issues is also growing in my country of birth: the United States (U.S.). Nevertheless, most of my earliest memories involved friendly interactions with fellow working-class Southerners of Old and/or New World ancestries. I was blissfully unaware that a new crest of an old wave of global financial crises was transforming my country in ways that would later

---

[i] I use the term 'nation' to refer to any collective defined by ancestry (and language). I use the term 'country' to refer any collective defined by government (and geography).

INTRODUCTION

impact my life enormously. Exacerbating said crises was the advent of another officially-endorsed drug epidemic; the rollout of the next phase of the post-chattel slavery, prison-industrial complex; and the public deployment of a now legendary, military counterinsurgency-surveillance project: the Internet.

A few early memories involved witnessing, alongside perplexed neighbors and family members, supranatural phenomena – some quite heart-warming, others quite insidious. Therefore, by the time I entered the first grade, I knew there was more to reality than meets the naked eye. Such insight helped me to grasp the many classroom songs and chapel lessons about DIVINE miracles, neighborly love, and righteous living. As I progressed up the grades, I became increasingly intrigued yet confused as to why teachers at this Southern Baptist administered school never discussed the organized crime, institutionalized racism, and generational poverty pervading the coal-rich region where we all lived. I quickly learned that it is best to discuss such issues among loved ones.

Enter Mom: a masterful cook [Brooks, 2000]; licensed beautician; elder-caretaker; one of the first descendants of U.S. chattel slaves to become a police-woman in the city where I grew up; and a Christian [ii] by life example. She often finished our countless discussions of the aforesaid issues as follows: "for we do not wrestle against flesh and blood, but against principalities, against powers, against the rulers of the darkness of this age, against spiritual hosts of wicked-ness [iii] in the heavenly places" [Ephesians 6:12]. Before I transitioned from the private school into a public high school, Mom began tenderly yet persistently

---

[ii] A 'Christian' is any person who professes to believe Jesus of Nazareth is THE CHRIST or, in addition, whose life reflects the faith of THE CHRIST [Matthew 7; Acts 11].

[iii] 'Wickedness' is anything of 'evil' effect. Evil is not an entity per se – but rather, it is a movement of the will away from the DIVINE Good [Seco and Maspero, 2010].

urging me to read Proverbs. I would truthfully answer as she left for work each evening: "no". Weeks passed before the sincerity in her eyes moved me to relent. Starting with one random verse of Proverbs every other night, I slowly moved on to one random chapter per night for several months. The GOD of Scripture is faithful. Indeed, from that time on, I was continually shown that there is an infinitely more sublime and daunting source of socio-spiritual authority at work against as well as above the 'world system' [iv] described in Mom's favorite quote.

Halfway into rigorous undergrad studies at the University of the South, I was blessed with global opportunities to independently research my favorite topics: spirituality, music, nature, and technological innovation. Hailing from small-town USA, I was amazed that I could always use my U.S. debit card to buy credits for some mobile phone, text newfound friends, then meet up somewhere to enjoy conversation and music – as if I had never left my neighborhood. I was all the more amazed by the profound diversity, splendor, and power of nature. Then there were the manifestations of spirituality I encountered; some inclusive and evident, others exclusive and secretive. It was humbling to behold just how much spirituality, music, nature, and technology affects our daily lives.

After graduation, I unexpectedly found a way to pursue a longtime interest in socio-behavioral science – as well as honor my computer science professors – by teaching high school mathematics and technology. Two years into that career path, a friendly Rotarian told me of an opportunity to become the first (and, at that time, only) fulltime student in a new master's program in security informatics at Indiana University. It was incredible to study topics like access

---

[iv] I use the phrase 'the world system' to refer to today's, age-old, global system of socio-spiritual authority characterized by death/decay – in a physical and nonphysical sense. See Wallerstein [2004] for a similar, yet non-spiritual treatment of the world system.

INTRODUCTION

control design, secure crypto-processors, and metamorphic virus detection. Yet topics like steganography, pervasive computing, and confidence tricks intrigued me the most – for those studies introduced me to the term 'social engineering'. Hungry to learn more, I took extra-departmental coursework in the school of mathematics and the school of business. Studying topics like government-corporate security engineering and applied cryptography satiated my hunger – slightly. So I designed an independent study course on propaganda theory and progressed from public relations to military science. I kept seeing references to, yet no rigorous explanations of, the term 'psychological warfare'. Frustrated, I took to the Internet; and found myself tangled in a web of conspiracy theories.

A new, post-graduation crest of an old wave of global financial crises only exacerbated things for me – along with millions of other young adults becoming entangled in webs of conspiracy theories while trying to understand why our circumstances were so dismal. Said theories offered us various culprits to blame; like intergovernmental organizations and secret social networks – all controlled by powerful elites. What made said theories so enticing is that they contained grains of truth. Indeed, I was a nonmember ambassador for a nongovernmental organization whose early members helped to create the United Nations (U.N.). This intergovernmental organization has access to a range of elite power centers globally; from courts and universities, to corporations and armies. Moreover, Initiation into secret social networks was a rite of passage for many generations of aspiring youths in the region where I grew up. Those networks with the most economic-political-religious power have connections to the ancient 'Mysteries'.

What made said theories so harmful is that they fostered not only envy of elites' power but also the seeking of blind revenge for elites' oppression of the weak. Yet the GOD of Scripture is faithful. For I was reminded that elites are ultimately fallible fellow humans who must someday depart the domain of the living; stripped of all their power. I was slowly untangled from the conspiracy theory web. Once liberated, I realized that I had undergone both psychological warfare (set within a larger context of social engineering) and 'spiritual warfare'.

Consider once again the U.N., which is at the forefront of 'sustainable development': a global campaign to avert nature's total destruction by steering both the physical (meaning population) and spiritual growth of humankind [Erdmann, 2005; Erdmann, 2009; Mosher, 2011]. Also consider the founder of one of the few nongovernmental organizations authorized to partake in the U.N.'s work wrote books under the guidance of influential suprahuman parties.[v] Such 'divinely-inspired' literature revealed the latest iteration (ending in 2025) of an ancient war to avert the world system's total destruction – by cultivating and reinforcing divinely-inspired conceptions of material, corporeal, mental, and spiritual empowerment among the global masses [Besant, 1919; Bailey, 1955; Bailey, 1957]. Then consider that the most influential such conceptions are intimately associated with the Mysteries. This secret social network consists of countless ideologically and morally unified, yet ethnically and culturally diverse women and men devoted to fulfilling the 'Great work'; by any means. Throughout world history, these and other influential human parties have variously identified the force energizing this grandiose endeavor as the 'Inner Sun', the 'gods', the 'Mother god(dess)', the 'Ancient Ones', the 'Shining Ones', etcetera [Pike, 1874; Billington, 1980; Ferguson, 1980; Hubbard, 1993; Leet, 1999; Peterson, 2003; Uždavinys, 2004; Monaghan, 2011; Uždavinys, 2011].

---

[v] I use the phrase 'influential *suprahuman* party' to refer to the noncorporeal constituents of the world system who – by way of their corporeal possessions/devotees; referred to as 'influential *human* parties' – relentlessly seek to sway the global masses' worldviews.

INTRODUCTION

Interestingly, Scripture addresses spiritual warfare, social engineering, and psychological warfare often. It identifies the ultimate originator of all three as a supranatural kingdom. Comprising this global system of socio-spiritual authority is a hierarchy of wicked suprahumans variously identified as powers, 'demons', 'elementals', 'the gods of the nations', 'the rulers of the world system', etcetera. Still more interestingly, the chief ruler is variously identified as a Shining One, 'the Serpent', 'the god of the world system', 'he who blinds the minds of humankind', 'a murderer from the beginning', etcetera. These entities' core message to humankind is referred to as 'the Lie'. Most interestingly of all, our embrace of – and, by extension, our complicity in – this prototypical, conscious-intentional falsehood is one of the main reasons why mass oppression and self-destruction has characterized life on the Earth from antediluvian times up to today [Genesis 3, 11; Deuteronomy 18; Psalm 82, 91; John 8, 12-16; 2 Corinthians 3-11; Galatians 4; Ephesians 1-6; Colossians 2; 2 Timothy 2-3].

Greatly expanding upon my article published in the *Journal of Information Warfare* [Davis, 2010], the present book is an introductory concept explication (and literature review) of worldview warfare. It uses as the guiding case in point the ancient, ongoing endeavor to fulfill the Great work. To that end, it explores some of this grandiose endeavor's patterns of development and execution across the following battlespaces:

- **Seven of culture's core components**: religion, art, politics, law, economics, philosophy, and science.

- **Seven of world history's most significant societies**: Egypt, Babylonia, Israel, Phoenicia, Greece, Rome, and the West.

- **Four of life's most transformative phenomena**: nature, technology, music, and spirituality.

St. Andrews community market

Victoria, Australia

NATURE

We acquire knowledge via physical (visual, audible, tactile, etcetera) and nonphysical (imaginal, spiritual) communication. Yet communication would be impossible if communicators and communicatees did not already share some preexisting verbal or nonverbal language. An individual's knowledge of that language empowers him or her to effectually operate in any collective (like a nation, kingdom, country, society, network, movement, or family) unified by it [Corbin, 1964; Eco, 1976; Lupyan, 2012]. Yet if that collective's constituents oppose its ruler, they can set its total destruction into motion – without any physical violence. Enter the *psychological*, *sociological*, and *pneumatological* dimensions of worldview warfare. Worldview warfare consists of cultivating and reinforcing worldviews that serve to either establish or maintain the rulership of a given party while simultaneously either eroding or precluding oppositional parties' rulership. Worldview encompasses a person's inner life (spiritual heart, mind) and outer life (words, deeds). Hence effectual worldview warfare is executed via offensively – as well as defensively – armed deeds, words, minds, and spiritual hearts [Bailey, 1957; Kitson, 1977; Simpson, 1994; Naugle, 2002; Hutchinson, 2006; Collins and Collins, 2020].

In the context of the war to fulfill the Great work, such militarization of inner and outer life is facilitated by what I refer to as 'Revolution'.[vi] Revolution is the *cultivation-reinforcement of a love* – the most consequential act a person can express or experience – *for ignorance, arrogance, and rebellion against the Truth*. Interestingly, such moral and spiritual darkness is familiar to every human who has ever lived under the shadow of the world system. Still more interestingly, what makes Revolution so seductive is that it exploits a critical vulnerability in all humankind: the desire for *material-corporeal self-gratification* and (especially) *psycho-spiritual self-glorification*. Most interestingly of all is Revolution's delivery mechanism. I refer to it as 'wicked deception'. Wicked deception is the inherently insidious, contradictory, and rousing communicative act of leading a person into believing that the Lie is the Truth.

[vi] I conceptualize my theory of worldview warfare as a (three) stepped triangle:

At the bottom step is the individual human – corporeally and mentally. I refer to the execution of worldview warfare here as 'psychological warfare' (Revolutionary, **checkered half**) or as 'psychological security' (*counter*-Revolutionary, **solid half**). At the middle step is the human collective – locally and globally. I refer to the execution of worldview warfare here as 'social engineering' (Revolutionary) or as 'social governance' (*counter*-Revolutionary). At the top step is the supranatural realm which transcends, yet is immanent in, the natural realm. I refer to the execution of worldview warfare here as 'spiritual warfare' – for it ultimately centers on either defending (Revolutionary) or endangering (*counter*-Revolutionary) the world system's interests.

NATURE

Knowledge of truth versus lies figures prominently among the set of core beliefs comprising a person's worldview. Consider semiotics: the sociological-psychological study of any natural or humanmade thing that can be substituted for, and thus counterfeit (lie about), another thing. Also consider demonology: the systematic-theological study of the suprahumans intimately associated with conscious-intentional falsehoods (lies) in general and the Lie in particular [Eco, 1976; Toorn et al., 1999]. Then consider that foundational to the set of core beliefs comprising any person's worldview is his or her conception of the self and the other. Throughout world history, the most consequential and divisive such conceptions have centered on DIVINITY.[vii] They can be summarized as:

- The self and the DIVINE are essentially **two**.
- The self and the DIVINE are essentially **one**.

These conceptions are mutually exclusive. It follows that anyone who embraces the false conception has embraced a lie. Furthermore, his or her own rejection (cognizantly or incognizantly) of the true conception ultimately renders them complicit in that lie [Eliade, 1962; Carson, 2010; Jones, 2010; Lachman, 2017].

Enter the fundamental element of Revolutionary worldview warfare so reliant upon the complicity of the TA (target audience; being a collective or an individual) that it is often portrayed as *inferior to physical violence*: propaganda. Indeed, propaganda's efficacy greatly depends on how much it resonates with the TA. The chief factors in such resonance include the propaganda's level of credibility, creativity, diffusion, and intensity. Moreover, physical violence – from social crises like political terrorism, to ecological crises like pandemics – can serve to enhance resonance. Highly intense, diffuse, creative, and credible propaganda is therefore often portrayed as *practically impossible to counteract*; let

---

[vii] I use the word 'DIVINITY' to refer to the quality of having DIVINE Personhood.

alone avoid [Bernays, 1928; Kitson, 1977; Sproule, 2001; Lipson and Binkely, 2004; Hutchinson, 2006; Snow and Taylor, 2006; Herf, 2009; Svensen, 2009; Bolt, 2011]. Most propaganda theories fall into one of those two extremes, yet Jacques Ellul's [1965] is a synthesis of both. For he contends that effectually resisting propaganda requires "an uncommon spiritual force or psychological education" [p.337]. Yet he concedes that "propaganda cannot create something out of nothing" [p.36]. For propaganda only builds upon the TA's preexisting spiritual heart issues (like deceitfulness, malicious pridefulness,[viii] unforgiveness, avariciousness, and lustfulness) as well as the TA's preexisting core beliefs.

Enter sub-propaganda; which exploits conditioned reflexes and expresses myths. Myths are emotive – even sacralized – narratives of how reality could or should be. Conditioned reflexes are expedient reactions – especially if lacking impulse control – to things that challenge those narratives. *Continual* exposure to sub-propaganda primes TAs *subconsciously* [ix] for action when the opportunity (whether genuine or contrived) arises. Enter active propaganda; all of which is ultimately integrative or agitative. Agitation propaganda cultivates-reinforces mass *strife* – often via simplistic, provocative, direct messaging. Integration propaganda cultivates-reinforces expedient forms of mass *solidarity* – often via indirect, evocative, subtle messaging. Active propaganda harnesses language's enigmatic power to not only reflect but also shape humankind's understanding and interpretation of reality [Ellul, 1965]. Consider how language can enable us to imagine certain ways of doing yet prevent us from imagining other ways.

[viii] One product of malicious pride is 'evil suspicion', wherein a TA is impugned without conclusive evidence [1 Timothy 6]. It plays a vital role in racial superiority theories.

[ix] Subconscious nonphysical or physical influences can greatly affect the mind and body [Estabrooks, 1943; Krippner, 1978; Bullock, 2004; Sathyanarayana et al., 2009].

NATURE

Also consider how by merely articulating a thing, we can invoke its (*imaginary*) presence-power as well as compensate for its (*actual*) nonexistence-impotence. Then consider how much what is spoken to us by other people – even ourselves – effects our deeds [Eco, 1976; Demeulemeester, 1994; Lupyan, 2012].

Propaganda comprises the rhetorical aspect of another fundamental element of Revolutionary worldview warfare: technique. The English word 'technique' is derived from the ancient Greek word 'technē': the skilled *Arts* of crafting physical – even nonphysical – things into useful products; namely, technologies. From that was derived the ancient Latin word 'ingenium': human – even divinely – inspired creativity, often in the context of warfare. From that was derived the English word 'engineering': a core product of the *Sciences*. Enter culture: the divinely-human inspired, nonphysical-physical 'codes of being and becoming' that we pass down through tradition [Ellul, 1980; Shepherd et al., 2006; Roochnik, 2007; Guliciuc and Guliciuc, 2010; Menezes de Carvalho, 2010; Wierzbicki, 2015; Collins and Collins, 2020]. On the one hand are the open codes comprising what I refer to as 'common culture'. Its core components are philosophy, religion, art, politics, law, economics, and science. On the other hand are the secret codes comprising what I refer to as 'Mystery culture'. Its core components are First Philosophy, Mystery Religion, the Sacred Arts & Sciences, Deep Politics, Sacred Law, and Shadow Economics.[x]

Technique is the manifestation of humankind's reliance upon culture (in general) and technology (in particular) to manage the existential, interpersonal, and informational insecurities inherent to life in the world system. Yet such good intentions for truth, love, and hope on our own terms are influenced by our own desires for self-gratification and self-glorification. Hence technique is

---

[x] I regard religion as the manifestation of spirituality.

characterized by secularization and syncretization. In syncretization, techno-cultural innovations redefine mutually exclusive phenomena; thereby confusing symbolic-functional distinctions. Chief such phenomena are life-death, evil-good, human-beast, male-female, and created-uncreated. In secularization, technocultural innovations replace orthodox sources of socio-spiritual authority with empowered alternatives. Chief such sources are GOD THE FATHER and HOLY Tradition [Ellul, 1980; Bainbridge, 1982; Ellul, 1986; Ascott, 2005].

Technique would be incapable of acclimating TAs to life in the world system if not for Revolutionary worldview warfare's fundamental element with the largest scope: the Zeitgeist. Encompassing the macrocosm of world history in general as well as the microcosm of any particular place-time; the Zeitgeist manifests as the collective unconscious and conscious. The collective conscious comprises the *shared avenues of human inspiration* – opinions, feelings, motives (perceived by the mind) – informing a collective's conduct and lifestyles. The collective unconscious comprises the *shared avenues of divine inspiration* – dreams, visions, promptings (perceived by the spiritual heart) – informing a collective's understanding and interpretation of reality [Schaeffer, 1976; Noll, 1994; Yates, 2001; Uždavinys, 2004; Ritenbaugh, 2005; Lachman, 2017].

Technology I

The notion of collective psycho-spiritual peace and material-corporeal prosperity within the world system – which is the fulfillment of the Great work on a macrocosmic level – has captivated countless ruling

NATURE

human parties throughout the ancient world. Yet before any such conception of utopia can exist in the physical realm, it must first take root in the soil of the local contemporary collective conscious; soil that constantly shifts atop the enduring global bedrock of the collective unconscious. So said parties employed skilled communicators (like orators, actors, musicians, painters, and writers) to cultivate-reinforce their conceptions of utopia among the ancient masses [Feen, 1996; Uždavinys, 2004; Török, 2011; Siniossoglou, 2011; Kelley, 2011; Gilchrest, 2013; Leprohon, 2014; Herrstrom, 2017; Seidenberg, 1961].

The quest for utopia has not ended. Yet ruling human parties of today have a significant advantage over their antecedents in antiquity: *electronic-digital* information technologies (EITs). EITs make mainstream as well as alternative radio, cinema, television, and video games possible. Programming these mass media's diverse content is a range of collectives; from government institutions and military intelligence services, to corporate institutions and organized crime [Bernays, 1928; Odom, 2003; Hutchinson, 2006; Dyer, 2016; Smith, 2017; Dyer, 2018]. Moreover, such media provide an avenue for regulating a range of mental events; from cognition and states of consciousness, to affection and conation. Enter 'cybernetics'. Cybernetics is rooted in the belief that all EITs as well as natural processes – even the Earth itself – can be precisely controlled by manipulating the singular substance ostensibly comprising physical reality: energy-information [Lilly, 1968; Delgado, 1969; Turkle, 2005; Schüll, 2012].

Governments and corporations have long been researching and developing various 'cybernetic ecosystems' (CESs); each comprised of clusters of EITs. Consider the Internet. This CES is facilitating levels of human interaction and interconnectivity not seen since a legendary monument was being erected in Babel. Also consider technological innovations like Nth Generation Mobile

Telecommunications and the Internet-of-Things. The latter is a major step towards wirelessly interconnecting most EITs on Earth. The former is all about the high capacity data transmission channels needed to make such connectivity possible. Both of these innovations are facilitating the proliferation of new CES products comprised of EIT clusters capable of automatically sensing – even interacting with – their surroundings. Some products have long been capable of monitoring – even modifying – data and voice transmissions in real-time. Then consider Eco-modification technologies and cryptocurrencies. The latter CESs are poised to transform the execution and regulation of commerce world-wide; via digital money. The former CESs are capable of modifying the Earth's lithosphere, atmosphere, and hydrosphere. Technology's social engineering-psychological warfare utility is colossal indeed [Brzeziński, 1978; Juda, 1978; Leigh, 2004; Thomas and Elliot, 2004; Snow and Taylor, 2006; MacArthur, 2007; Sioshansi, 2011; Levine, 2018; Daskalakis and Georgitseas, 2020].

Universal Physis

Nature is like unto a universal language that transcends biology – even corporeality. Diverse societies throughout world history have held such a conception of nature; like the 'physis' of the Greeks [Blainey, 1975; Welch, 1987; Pontifical Academy of Sciences, 2016]. Still more interestingly, such conceptions persist to today. The Ecological movement is a compelling case in point. On the surface, this global movement provides a creative means of bringing awareness to, as well as building consensus around,

NATURE

a growing number of genuine – even counterfeit – ecological crises. On a deeper level, it provides a means of advancing the endeavor to fulfill the Great work.

Enter the debate about who or what is ultimately to blame for the global intensification of ecological crises. Most positions in this increasingly polarizing debate place the blame either upon humankind in particular or upon nature in general. Adherents of the latter view largely attribute the global intensification of ecological crises to the normal operation of ancient, natural phenomena that the Sciences cannot explain. Thus they often accuse the other side of promoting economically as well as politically expedient pseudoscience [Lehr, 1992; Dewar, 1995; Gornitz, 2009; Ball, 2014; Darwall, 2017]. Adherents of the former view largely attribute the global intensification of ecological crises to modern humankind's abnormal impact on the Earth – an impact exacerbated by rising population levels, especially in developing countries. Thus they often accuse the other side of promoting scientifically-discredited conceptions of humankind as well as nature [Meadows et al., 2004; Diethelm and McKee, 2009; Berg, 2017].

Figuring prominently on both sides is a diverse school of philosophy that is articulated in (largely Westernized) terms of nature in particular and physical reality in general: Scientism. Scientism asserts that *all phenomena within the natural realm are best understood – and can even be controlled – via the Sciences.* Two of the most prominent forms of Scientism are Physicalism and Radical Empiricism. The latter centers on the human sensory organs, portraying them as the only avenues for acquiring knowledge suitable for corporeal survival. The former centers on physical reality in general and nature in particular, portraying them as the only verifiable bases of truth [Armytage, 1965; Voegelin, 1968; Billington, 1980; Dupré, 1993; Guénon, 2004; Collins and Collins, 2006]. Nature's psychological warfare-social engineering utility is colossal indeed.

A core component of science is physics. Physics is divided into quantum mechanics and general relativity. One of the most popular attempts to unify both is string theory. It posits that reality is ultimately comprised of countless, tiny strings of energy-information. A string's fundamental frequency informs the fundamental characteristics of whatever it constitutes at a given time. Hence string theory is often explained in musical terms [Kaku, 1999; Cooley, 2005]. Yet explications of reality in terms of musical vibrations – and conceptions of reality as being comprised of a singular substance – are far from new [Rudhyar, 1982; Christensen, 2002; Taylor, 2007; Marin, 2009; Siniossoglou, 2011].

The global Ecological movement has contributed immensely to the modern revival of the ancient Animist assertion that the life principle (soul) of every natural entity – including humans and even minerals – is interconnected via the 'World Soul'. Thus the global intensification of ecological crises signifies that we are out of harmony with 'Mother Earth' [Capra, 1983; Cooper, 2006; Svensen, 2009]. This and other tenets of Ecological philosophy (known as 'Ecosophy') inform a growing amount of 'Eco-edutainment' mass media; from websites and movies, to magazines and music [Tedesko, 1992; St. John, 2006; Naess, 2007; Murray and Heumann, 2009; Maxwell, 2010; Pedelty, 2012].

Deeper Levels

The worldview underlying *Animism* asserts that the singular, amoral, impersonal, supranatural World Soul generated and sustains every natural entity. Such conflations of the self with the DIVINE also exist

NATURE

within *Atheism* and the *Mysteries* [Eliade, 1962; Cooper, 2006; Hedges, 2009; Uždavinys, 2011]. The worldview underlying *Scriptural Judaism* and *Scriptural Christianity* asserts that all entities – natural or supranatural – are sustained by a Personal, HOLY (perfectly righteous), Triune uncreated CREATOR [Ware, 1995; Barker, 2010; Ellingsen, 2015]. While highly divergent overall, several points of similarity exist between these two worldviews – points long exploited by countless influential human parties worldwide for purposes of secularization-syncretization [Hooper, 1906; Bruteau, 1974; Schuon, 1984; Scholem, 1990; Smith, 1995; Arnold, 1996; Rosenthal, 1997; Vasilev, 2014; Harari, 2017].

One of the most significant points of similarity centers on the intimate association between knowledge and (corporeal as well as spiritual) survival. Enter Supranaturalism. This diverse school of First Philosophy asserts that *all phenomena within and beyond the natural realm are best understood – and can even be controlled – via the Sacred Arts & Sciences.* Fundamental to the understanding aspect is 'divine revelation': otherwise hidden knowledge imparted to humans by suprahuman entities variously referred to as powers, Ancient Ones, Shining Ones, etcetera. Fundamental to the control aspect is technology devised from such divine inspiration [Underhill, 1930; Brann, 1999; Shaw, 2003; Uždavinys, 2011]. Both factors require devoted training, the surface levels of which involve cultivating a deeper awareness of the supranatural realm. Deeper levels involve reinforcing intimate relationships with said realm's inhabitants: the aforesaid suprahuman entities renowned throughout world history for nonphysically (and even physically) empowering as well as tormenting humans [Michaelsen, 1982; Charet, 1993; Nicholson, 1987; Steiner, 2009; Gallagher, 2020; Virtue, 2020].

A core component of the Sacred Arts & Sciences is alchemy. Alchemy centers on the microcosmic and macrocosmic transmutation of soul. The latter involves humankind's collective participation in the Great work throughout world history. The former involves individuals' life-long journey along at least one of the four interrelated 'pathways of the gods' [Besant, 1919; Bruteau, 1974; Franz, 1979; Goodman, 1980; Scholem, 1991; Hubbard, 1993; Kripal, 1998; Puett, 2002; Shaw, 2003; Uždavinys, 2008; Finnegan, 2013; Lampe, 2014]:

- **Intellectual**: acquiring-applying divine revelation with the intention of imaginally unifying the self and the DIVINE, in essence.

- **Liturgical**: performing an extended series of rituals (a working) with the intention of manifesting suprahuman presence-power in the natural realm.

- **Experiential**: entering into an altered state of consciousness (ASC) with the intention of navigating, as well as transacting in, the supranatural realm.

- **Behavioral**: living a hedonistic or ascetic lifestyle with the intention of meriting some supranatural advantage in life or after death (divine favor).

Via adherence to a given path – especially the intellectual and liturgical paths – the journeyer ostensibly rises ever-closer to a GOD-like status. The *few* who have the sociocultural capital to endure such a journey effectively earn the right to rule (in any manner they see fit) over the *many* who do not [Salk, 1973; Wilmshurst, 1980; Fishwick, 1992; Puett, 2002; Dunand and Zivie-Coche, 2005]. These parties' most effectual means of establishing or maintaining rule (while eroding or precluding opponents' rule) have long centered on exploiting the communicativeness of **money, blood**, and **DIVINITY**.

The roots of such Revolutionary worldview warfare in the West trace back to ancient Greece. Ruling Greek parties often deliberated the soul's relationship to the blood. They intimately associated blood with the body and the mind; on not only the individual level but also the collective level. This informed their endorsement of, and participation in, a range of blood workings; from human

NATURE

sacrifice and neglect of 'useless eaters', to pederastic education for upper-caste boys and prevalent ASC-inducing substance abuse. These workings comprised some of the first sustainable development measures in the West [Galton, 1998; Percy, 1998; Hillman, 2014; Andreadaki-Vlazaki, 2015; Boylan, 2015; Lavelle, 2016]. Enter Hellenic philosophy: the divinely-inspired, 'verbal drug' that played a vital role in rationalizing, romanticizing, and legitimizing the aforesaid measures. Premised upon Egyptian First Philosophy, Hellenic philosophy was intended as a contemplative lifestyle whereby Greeks could attain Illumination; ostensibly leading to total union of the self and the DIVINE (in essence) – which is the fulfillment of the Great work on a microcosmic level. Building upon this proto-evolutionary theory of self-salvation was Hellenic 'environmentalism': a proto-racial superiority theory portraying most non-Hellenes as the beast-like products of inferior mental makeups and ecological conditions. Such pseudo-scientific theories persist in today's Ecosophy [Davis, 1986; Feen, 1996; Isaac, 2006; Uždavinys, 2004; Naess, 2007; Hinrichsen, 2010; Herrstrom, 2017].

The Romans innovated upon Hellenic Revolutionary worldview warfare. Sports provided upper-caste Romans with an intoxicating avenue for chasing after the prototypical approximation of Immortality, fame – often in the context of shedding lower-caste people's blood. For physical violence was employed throughout this immense empire as a means of both personal and political crisis resolution. Exacerbating things, ASC-inducing substance abuse pervaded every caste. Greatly facilitating the herculean task of uniting this diverse empire socioculturally was 'civil religion'. This culture-encoding technology introduced the worship of Pharaoh-like Caesars into the Hellenic tradition of worshipping the gods and the ancestors [Fishwick, 1992; Futrell, 1997; Hillman, 2014]. Consider Caesarea Paneas: the Western Asian town situated around a watery cave at the base of a legendary mountain. Even before Roman times, this town was known in the *Mysteries* as a key crossroads of the netherworldly domain of the ancestors/dead, the earthly domain of the living, and the heavenly domain of the gods. Also consider that this town had long been intimately associated with the Egypto-Hellenic god Pan; whose worship was characterized by lucid (often musical) psychoses, violent pansexual love, and ecocidal nature-worship. This same dark-light dynamic persists in today's global Ecological movement. For it often portrays humankind as a 'cancer' on the planet; yet incessantly calls for humankind to heal the planet [Hughes, 2014; Taylor, 2010; Wilson, 2004; Meadows et al., 2004; Borgeaud, 1988]. Then consider that Scripture identifies Caesarea Paneas as the location where an Israelite named Yeshua first publicly ordained *Scriptural Christianity* – near a gleaming temple dedicated to a Caesar revered in life as the 'son of god' [Matthew 16; Mark 8; Luke 9; Peppard, 2011].

The Christianized European kingdoms arising out of the Western Roman Empire's ashes innovated upon Greco-Roman Revolutionary worldview warfare. This Christian Empire (in keeping with its chief rival, the Islamic Empire) embraced various alchemical practices [Peters, 1998; Kelley, 2011; Uždavinys, 2011; Heather, 2014]. Consider the perpetuation of Animist blood workings in the form of 'holy war' and allegorized pederasty as well as the prevalent abuse of ASC-inducing substances; like cannabis, opium, and alcohol [Murray and Roscoe, 1997; Felsen and Kalaitzidis, 2005; Engel, 2006; France, 2006; Hillman, 2014; Rosenthal, 2015]. Also consider the imaginal workings carried out via divinely-inspired literature, paintings, temples, music, dramas, pageants, and storytelling [Farmer, 1925; Lethaby, 1981; Roob, 2006]. Then consider that such workings persist in the civil religion which is today's global Ecological movement via Eco-edutainment, Eco-pansexuality, and population control [St.

NATURE

John, 2006; Mosher, 2011; Pedelty, 2012; Sprinkle and Stephens, 2017]. Also persisting today (albeit, on a far greater scale) are pandemics, famines, economic disparity, and political terrorism [Herlihy, 1997; Blyth, 2013; Waal, 2017].

Ruling Mystery parties harnessed the Revolutionary worldview warfare utility of those social and ecological crises to portray the GOD of Scriptural Christianity as an unreliable and illegitimate FATHER. For the psychological security of *DIVINE Salvation* was the key obstacle to the Renaissance promotion of '*self-salvation* via (hidden) knowledge of nature'. So said parties administered various 'scientific priesthoods' whereby the upper-classes could journey along the intellectual pathway of the gods. Said priesthoods innovated upon medieval ruling Christian parties' methods of thought control. They officially censored rigorous theological and scientific confutations of Scientism. They minimized Scriptural Christianity's (namely, the Northern African and Celtic Church's) contributions to scientific progress in early medieval Europe [Armytage, 1965; Vickers, 1986; Collins and Collins, 2006; Oden, 2007]. Moreover, ruling Mystery parties administered various vocations and secret networks whereby the masses could merit divine favor; like spreading Eurocentric civil religions or otherwise conquering lands far from Europe. These conquests would situate Western Europe at the epicenter of an immensely lucrative, global flow of raw materials, money, drugs, arms, and slaves [Yates, 2001; Felsen and Kalaitzidis, 2005; Bridenthal, 2013; Schuchard, 2013; Bogdan and Snoek, 2014].

Moorish Muslims purchased countless indigenous Europeans from European human traffickers while executing the Islamic Empire's utopian agendas; and thereby innovated upon trans-Mediterranean slave networks existing long prior to Islam. European Christians were now purchasing countless indigenous Africans from African human traffickers while executing the Western Christian

Empire's utopian agendas; and thereby innovated upon trans-Saharan slave networks existing long prior to Christianity [Fynn-Paul, 2017; Milton, 2012; Pestana, 2011; Lovejoy, 2011; Karras, 1988]. Influential human parties at work within the Church of the West contributed enormously to the formation and perpetuation of this trans-Atlantic (and later, trans-Pacific) slave network; by devising de-humanizing – even anti-human – conceptions of the other. These agitative conceptions not only went on to permeate the collective conscious of the West (with most ruling Western parties' complicity) but also put a modern, Scientistic face on various ancient, Supranaturalistic ideas. Consider how said parties devised pseudoscientific theories of racial superiority; by repackaging the *Islamic* ideology that black skin – especially in regard to most [xi] people of African ancestry – signifies divine disfavor and perpetual servitude. Also consider how said parties refined *Hellenic* environmentalism to portray most non-Europeans as the beast-like products of inferior technocultural, ecological, and biological circumstances. Then consider how said parties devised pseudoscientific theories of evolution; by extrapolating tenets of *Egyptian* First Philosophy into Physical-ism. Such is how the integrative notion of the morally,[xii] mentally, corporeally, and materially superior 'god-Men' persists from antediluvian times up to today [Taha, 2005; Isaac, 2006; Uždavinys, 2008; Beer, 2018; Collins and Collins, 2020; Goldenberg, 2017]. Germany and the

---

[xi] Indeed, there were exceptions. The black-skinned Africans who the Islamic Empire could not conquer, the Aksumites of Ethiopia, are a compelling case in point. As one of world history's first Christianized kingdoms, Aksum gave rise to the same Eastern African Church still thriving today [Bowersock, 2013; Getahun and Kassu, 2014].

[xii] Moral superiority here calls for 'transcending' the conscience. Scripturally speaking, persons with such a 'seared' conscience pursue their wicked desires with no moral compunction for destructive effects upon other people [Romans 1-2; 2 Timothy 4].

NATURE

U.S. are compelling cases in point.

Ruling U.S. parties have, since before the country's founding, harnessed the Revolutionary worldview warfare utility of racial superiority theories to keep the multi-ethnic U.S. masses from uniting in opposition against them. For such culture-encoding technologies served to legitimize the introduction of a novel imaginal-legal collective: the 'white race'. Many TAs of European ancestry who embraced whiteness were rewarded with both economic inclusion and political advantage. Indeed, the rules for inclusion varied as per expediency. For some TAs of non-European ancestry accessed whiteness early – often via Initiation into Atlantic Mystery Religion [Hall, 1951; Walker, 2010; Battalora, 2015; Horne, 2016]. Many TAs of indigenous or non-Nordic ancestry were excluded from whiteness – often until they embraced racial superiority theories. For said theories served to rationalize the insatiably lucrative enslavement of most black-skinned peoples, portraying them as *naturally depraved*; while romanticizing the insatiably lucrative expropriation of resources from most indigenous peoples, portraying them as mere *ecologically-sophisticated savages* [Krauthamer, 2013; Ignatiev, 2012; Hinrichsen, 2010; Berreman, 1960]. Ruling U.S. parties went on to use Britain's racialized theories of evolution to repackage their grandiose endeavor to establish a master-race utopia as an *evolutionary mandate* to 'un-wild' (via the domestication of plants, beasts, and humans) then 're-wild' a new Eden. Said rulers (females as well as males) exhibited *no moral compunction* for the countless ritualized acts of violence [xiii] they endorsed and carried out against

---

[xiii] Indeed, there were blood workings. Labelled by upper-class whites as 'landless trash', countless lower-class whites were exploited as cannon fodder. Many whites of all classes resolved personal conflicts via 'honor murders' and satisfied 'divine debts' via indigenous massacres. Many indigenous peoples massacred whites. Some indigenous

fellow humans of indigenous, African, and/or European ancestry [Zafirovski, 2009; Isenberg, 2016; Powell, 2016; Curry, 2017; Jones-Rogers, 2019].

Pre-Nazi Germany saw the rise of a form of European environmentalism that played a vital role in the birth of the Ecological movement. This 'Blood and Soil' ideology centered on both ecological and social sustainability. Greatly contributing to its mass acceptance was its resonance with traditional Germanic Animism – as well as the spread of 'New Age': a British-U.S. innovation upon Nordic, Hellenic, and Hebraic Supranaturalism. Blood and Soil became a core component of a civil religion led by a Caesar-like führer: Nazism [Chamberlain, 1911; Webb, 1976; Biehl and Staudenmaier, 1995; Jacobsen, 2005; Chapoutot, 2016; Weikart, 2016; Hanegraaff, 2018]. Vast support from Atlantic – namely, U.S. and British – Shadow Economics and Deep Politics further helped the Nazis establish rulership over Germany. Ruling Nazi parties went on to enact 'Eco-legislation' facilitating unprecedented government-corporate *protection* of sacred lands (via military expropriation of natural resources beyond Germany's borders) as well as *management* of profane populations (via the humanitarian eradication of useless eaters) [Sutton, 1976; Billington, 1980; Weindling, 1989; Musser, 2010; Mosher, 2011; Waal, 2017; Whitman, 2017; Tate, 2019].

After World War II, ruling North and South American parties welcomed into their countries not only Nazi environmentalism but also Nazi Initiates of

---

peoples perpetuated chattel slavery. Many indigenous youths later endured torturous 'reeducation camps' [Wyatt-Brown, 1982; Krauthamer, 2013; Isenberg, 2016; Juster, 2016; Churchill, 2004]. Some blacks perpetuated chattel slavery. Countless blacks endured rape (often as children), medical experimentation plantations, and 'family-friendly lynching festivals' [Dray, 2002; Koger, 2012; Woodard, 2014; Kenny, 2015].

NATURE

high degree [xiv] – thereby facilitating the birth of the Ecological movement on a global level. Consider the increase in U.S. theological-academic portrayals of population control as salvific and Scriptural nature-stewardship as patronizing. Also consider the proliferation of U.S. government-corporate endorsed, Eco-edutainment about a population explosion of *dysgenic* plant, beast, and human invaders [White, 1967; Manning, 1981; Rosen, 2004; Bellin, 2009]. Having already been exposed to decades of similar messaging, the U.S. masses went on to widely support the enactment of Eco-legislation later blamed for facilitating countless foreign and domestic atrocities. Then consider that the victims often either lived on lands containing natural resources long coveted by ruling U.S. parties – or occupied some demographic long deemed by said parties to be full of useless eaters [Dewar, 1995; Cramer, 1998; Perkins, 2004; Isenberg, 2016; Powell, 2016; Darwall, 2017; Immerwahr, 2019; Molitch-Hou, 2019].

Most interestingly of all, world history shows that one of the most potent avenues for cultivating-reinforcing such expedient conceptions of the self and the other is music.

[xiv] A main reason why said Initiates were welcomed was because of their Revolutionary worldview warfare expertise. For they had effectively modernized it [Simpson, 1994].

The highlight of my first Disney experience

Tokyo, Japan

MUSIC

The phenomenon of music encompasses the following six domains [Sachs 1940; Brown, 1970; Hopkin, 1996; Murray and Wilson, 2004; Roederer, 2008; Meyer and Hansen, 2009; George-Graves, 2015]:

- **Words** (audible to visual): singing, instrumentation, dancing, and writing.
- **Numbers** (spatial to temporal): position, frequency, repetition, and rhythm.
- **Imagery** (mental to physical): imagination, body gestures, symbol form, and artifact form.

Music affects human learning, remembering, and decision-making on an intimate level. Furthermore, music comprises one of the most powerful avenues for inducing an ASC [Farmer, 1925; Scott, 1935; Husch, 1984; d'Olivet and Godwin, 1987; Ballam, 1994; Cross, 2001; Liikkanen, 2008]. Individuals who enter into an ASC – even a *short-term* ASC, if intense enough – may experience *long-term* changes in their inner and outer lives. And during group-ASCs, each person in the group may experience shared sensory stimuli, emotions, thoughts, and situations [Estabrooks, 1943; Hollingshead, 1974; Goodman, 1980; Grof, 1989; Strassman, 2000]. Music's psychological warfare utility is colossal indeed.

Inquiries into music's effect upon the mind and the body played a vital role in the creation of psychophysics: a field often credited with shifting psychology away from the domains of religion and philosophy – into the domain of science. *Outer* psychophysics focuses on relationships between stimuli (physics) and neural activity (physiology). *Inner* psychophysics focuses on relationships between electrical signals (physics) and brain functions (physiology) – as well as between corporeal (physiology) and cognitive-affective-conative (mental) events [Geissler et al., 1992; Kimble et al., 1996; Cacioppo et al., 2007].

Music has long provided fertile grounds for psychophysical research. For example, whole-body applications of musical stimuli containing ultrasound (approximately from 50 to 100 kHz) can elicit measurable, reproducible changes in attitude and behavior. Every major musical instrument group – especially the percussion group – contains members capable of generating ultrasound [Boyk, 2000; Oohashi et al., 2000; Oohashi et al., 2006]. Whole-body applications of infrasound (approximately 20 Hz or less) can elicit measurable, reproducible psychological and physiological changes. Among the most popular of the many musical instruments capable of generating infrasound is bass percussion [Johnson, 1980; Hope, 2008; Leventhall, 2009].

Psychophysical research also shows that repetitive, rhythmic musical stimuli can elicit remarkable psychological and physiological changes. Music rhythms similar in intensity but dissimilar in tempo are capable of inducing corresponding brain rhythms – even altering brain functioning. Then there is ambiance (the mood associated with a given sound, person, location, situation, or artifact). Ambiance can significantly amplify the psychophysical effects of

MUSIC

rhythm, repetition, and frequency. Indeed, music affects the mind and the body in ways that Scientistic terms cannot rigorous explain [Rudhyar, 1982; Husch, 1984; Ballam, 1994; Diamond, 1997; Oohashi et al., 2000; Winkelman, 2003].

Inquiry into the relationships between the mind and the body is a unifying theme among a range of scientific fields; from psychophysics and quantum mechanics, to ecology and cryptography. However, many of these Scientistic fields' luminaries were deeply influenced by Supranaturalism [Vickers, 1986; Brann, 1999; Jacobsen, 2005; Marin, 2009]. Consider the field of psychology; which is divided into four orientations. The humanistic-existential orientation focuses on self-understanding and self-awareness. The cognitive-behavioral orientation focuses on collective thought and collective deed. The psycho-dynamic orientation focuses on conscious and subconscious self-motivation. The transpersonal orientation focuses on the collective unconscious, a psycho-socio-spiritual dynamic that arguably gives it the most significant Revolutionary worldview warfare utility of all four. Indeed, a range of socialization programs increasingly espouse a transpersonal orientation; from personal and professional development for adults, to outreach and education services for youths [Adeney, 1981; Muller, 1982; Grof, 1989; Charet, 1993; Kimble et al., 1996; Hall and Edwards, 2002; Semetsky, 2012; Friedman and Hartelius, 2013].

Also consider transpersonal psychotherapy's 'active imagination', in which therapists utilize ASCs to help patients tap into the collective unconscious and translate the subsequent divine inspiration into observable performance. Hence two of the most popular avenues for active imagination are music imaging and dance therapy, both of which heavily rely upon musically-driven ASCs. Then consider that, during active imagination sessions, patients often *involuntary* carry out repetitive, rhythmic communicative acts; like gesturing, speaking, or writing [Tilly, 1977; Franz, 1979; Chodorow, 1991; Beebe and Wyatt, 2009]. For active imagination puts a pseudoscientific face on divination: the *Sacred Art & Science* of acquiring divine revelation. Divination is intimately associated with 'possession'.[xv] Possession is an enigmatic phenomenon characterized by a range of invasive effects; from *involuntary* body movements and mental imagery, to *compulsive* lifestyles and conduct [Gallagher, 2020; Jung and Shamdasani, 2009; Noll, 1997; Charet, 1993]. Furthermore, throughout world history, facilitation of – and even total liberation from [VanderKam and Flint, 2002] – possession has been intimately associated with music [Rouget, 1985; Shaw, 2003; Walter and Fridman, 2004; Filan and Kaldera, 2009; Staemmler, 2009; Till, 2009].

Technology II

Extensive archeohistorical evidence tracing back to the time of Babel shows that every major musical instrument group – except those in which electrical signals are the main means of generating sound – was already in existence by then. Moreover, said instruments not only emerged quite

---

[xv] Scientism denies the spiritual reality of possession, relegating it to brain functioning primarily. Supranaturalism accepts its spiritual reality, asserting it can be beneficial or harmful. Further complicating things, possession has long been counterfeited by countless fraudsters – and even conflated with mental disorders; like psychosis. HOLY Tradition refers to it as 'demonization': a term describing situations wherein at least one wicked suprahuman (having already gained some avenue into a human's life) exercises various degrees of control over his or her mind and/or body; up to the point of complete control [Krippner, 1978; Goodman, 1980; Arnold, 1992; VanderKam and Flint, 2002; Filan and Kaldera, 2009; Gallagher, 2020; Heiser, 2020].

MUSIC

suddenly in Mesopotamia but also were quite sophisticated in design. Similarly sophisticated musical instruments would later emerge within diverse societies worldwide [Sachs, 1943; Duchesne-Guillemin, 1981; Blades, 1992]. We are now exposed to more music than ever – largely due to the global proliferation of musical EITs; like portable digital audio devices [Bovermann et al., 2016; Schouhamer, 2010; Liikkanen, 2008]. Then there are the musical and non-musical EIT clusters that comprise a growing number of musical CESs.

A core component of musical CESs is loudspeaker innovation; like sub-woofers and phased array speakers. Some of the latter can transform *ultrasound* into directional, *audible sound*. Many of the former can generate *infrasound* and plays a vital role in a range of settings; from bars/clubs and personal vehicles, to massive stadiums and home theaters [Pompei, 2002; Hope, 2008]. Another core component of musical CESs is music-as-a-service: an aspect of cloud computing. Cloud computing is enabled by high capacity datacenters and the Internet. Services include automated music prescription based upon musical EIT users' activities, preferences, mood, etcetera [Celma, 2010; Morris, 2011].

Interestingly, the greatest component of musical CESs is neither digital nor electronic. This human component not only enhances the musical EIT user experience the most but also increases the likelihood of said TAs exhibiting the most expedient attitudes and behaviors. Enter para-social interaction: a psycho-socio-spiritual dynamic that the growing numbers of fatherless youths today are especially vulnerable to. For in their earnest search for identity and affirmation, fatherless youths throughout the West are increasingly gravitating towards substitutive – in this instance, Internet-mediated – relationships with famous musicians who speak to their insecurities. Exacerbating things, these TAs are seldom presented with genuine representations of the musicians they esteem so highly – musicians who have limited ability (and often no desire) to form caring, accountable relationships with them [Lewis, 2009; Krause et al., 2018].

Universal Mousikē

Music is like unto a universal language that transcends the human senses – even the natural realm. Diverse societies throughout world history have held such a conception of music; like the 'mousikē' of the Greeks [Ellfeldt, 1976; d'Olivet and Godwin, 1987; Murray and Wilson, 2004; Walter and Fridman, 2004; St. John, 2006]. Still more interestingly, such conceptions persist to today. The Hip Hop movement is a compelling case in point. On the surface, this global movement provides a creative means of building consensus around, as well as bringing awareness to, a growing number of genuine – even counterfeit – social crises. On a deeper level, it provides a means of advancing the endeavor to fulfill the Great work.

Ruling human parties are supposed to address the prosperity and peace of the whole of their societies. Yet said parties often only address the prosperity and peace of the powerful few while neglecting the weak masses. Enter the Hip Hop movement: a sociocultural innovation upon the legendary tradition of civil resistance in the West [Wolin, 2009; Daube, 2011]. Lower-caste descendants of U.S. chattel slaves and lower-class Caribbean immigrants to the U.S. started it. Some just wanted a fun escape from the gloom of poverty. Others wanted to resist the government officials who enact policies largely benefiting U.S. upper-classes, as well as the impersonal corporate forces long dominating U.S. political

MUSIC

processes. Over the years, this movement would be joined by growing numbers of demographically-diverse, anti-establishment (as well as pro-establishment) collectives worldwide – for the purpose of non-violent (as well as violent), civil engagement. Those collectives range from transnational advocacy groups and entertainment companies, to foreign policy/intelligence operators and religious institutions. Music's social engineering utility is colossal indeed [Toop, 1984; Donalson, 2007; Baker, 2011; Nitzsche and Grünzweig, 2013; Katz, 2019].

Westernized Countries: The U.N. had long played a vital role in the global implementation of ruling Atlantic – namely U.S. – parties' utopian agendas [Erdmann, 2005]. To help cultivate-reinforce the worldviews most expedient to said parties' crisis resolution (military) efforts, the U.N.'s Department of Peacekeeping Operations briefly partnered with U.S.-based, Afropop World-wide. This Peabody award-winning, nonprofit is skilled at targeting African-influenced music to youths in developing countries – even low-intensity conflict zones – undergoing Westernization [Scheuerman, 2008; Afropop Worldwide, 2016]. To help cultivate-reinforce the worldviews most expedient to ruling Atlantic parties' sustainable development (humanitarian) efforts, the U.N.'s Department of Public Information briefly partnered with U.S.-based, Music Television Networks. This Emmy award-winning pioneer in the global music industry later featured a famous, 'conscious-gangster' Hip Hop Initiate in some online Eco-edutainment media content targeting youths in Westernized, developed countries [Music Television Networks, 2007; UNESCO, 2009].

Oceania: Socio-ecological justice is an increasingly strong undercurrent in the global Hip Hop movement. Enter Combat Wombat: the self-identified 'troupe of Hip Hop punks' based in Melbourne, Australia.[xvi] They have long waged a 'social justice terra-ism' campaign to nonviolently pursue their political agendas outside of mainstream political processes; an approach in keeping with other transnational advocacy groups worldwide. Indeed, on foreign fronts, they championed the 'United Struggle Project' and promoted the 'Freedom Flotilla to West Papua'. On domestic fronts, they facilitated collaborations between nonindigenous and indigenous youths on sustainable development. They also exposed instances of government-corporate collusion in ventures that generated massive amounts of pollution – something indigenous peoples there have long been subjected to [Mitchell, 2006; St. John, 2010; Minestrelli, 2017].

Caribbean: Ruling U.S. parties have long executed nonphysical warfare against European rivals for control over the strategically-placed island of Haiti. The U.S.-endorsed rule of Rafael Trujillo over Haiti's eastern two-thirds (one of the most 'racially-intermixed' countries in modern history, in terms of European/indigenous/African ancestries) is a compelling case in point. For Trujillo harnessed the racial superiority theories of the U.S., the Dominican Republic's enduring military-colonial ethos, the Spanish nationalism of Francisco Franco, and Roman Catholicism's numerous points of similarity with African Mystery Religion. African-influenced music figured prominently in this psycho-socio-spiritual dynamic that persists to today [Austerlitz, 1997; Sidanius et al., 2001; Peguero, 2004; Derby, 2009; Horne, 2015]. Indeed, while strolling along Santo Domingo's El Conde, I have often observed fabric canvases, concrete walls, and

---

[xvi] I first learned about Combat Wombat through one of its members, 'Elf Tranzporter'. During a chance encounter on a Monash campus, this U.S. rapper invited me to a small lounge in downtown Melbourne for a freestyle rap battle with him and a couple Aussie rappers. It turned out to be my first public rap performance.

MUSIC

metal shutters serve as U.S.-flavored tributes to Dominican nationalism. In Colonial Zone discos, I have often witnessed Dominican-based merengue – the instrumentation of which shifted from Old World style to New World style during the late 20th century – and U.S.-based Hip Hop dance merge to reveal the ancient African roots of both. And while driving throughout the country, whether listening to radio stations or passing by countless 'colmados', I have often heard the latest beats and rhymes of U.S. as well as Dominican artists.

Europe: Seeking justice (as well as revenge) for Franco's decades-long campaign of genocide against the Basque, a secret army within their national liberation campaign has long executed war against the Spanish establishment. A legendary nonphysical aspect of this ongoing war involved this secret army's transition from punk rock to conscious-gangster Hip Hop [Llera et al., 1993; Mitchell, 2001]. Yet the Basque are not alone in harnessing the global Hip Hop movement to engage in civil resistance against ruling Spanish parties. Consider Salamanca, where I would sometimes observe circular arrangements of young Castilian locals (often wearing punk rock fashion) and young African migrants engaging in lively 'rap battles' at public parks. One evening, some of them called me into battle. A few others later called me into battle at their private homes. Also consider that the chief target of their verbal assaults was corrupt Spanish religious, political, and economic institutions. Then consider that such anti-establishment sentiment will only intensify as the financial austerity policies of governments worldwide continue disaffecting their youth populations.

Asia: The late 20th-century collapse of Asia's economic bubble contributed greatly to the global Hip Hop movement's entrance into Japan. Young students must navigate the dual pressures of Westernization and Japanese nationalism. Young workers must balance the despair of economic disparity with their hopes for future success. Hip Hop has come to shape as well as reflect both groups' conceptions of what constitutes genuine success and acceptable self-expression [Mitchell, 2001; Matsumoto, 2002; Condry, 2006]. I caught glimpses of this psychosocial dynamic one afternoon while briefly observing some middle school-aged youths engage in a Hip Hop dance battle near the entrance of a Tokyo subway station. They briefly stopped battling to observe me: a very tall representative of Hip Hop's ethnic root. And in Tokyo's Shibuya ward, I would constantly lose count of all the young adults wearing expensive, foreign-branded Hip Hop fashion. I chose to purchase a more affordable, and far more stylish, domestic-branded piece.

North America: Hyper-consumerism, stylish nihilism, and youth (labor and sexual) victimization characterize the global Hip Hop movement's dark side. It is chiefly expressed through gangster Hip Hop. The western U.S. originators of gangster Hip Hop were heavily influenced by punk rock ideology; which puts a Modernist face on Renaissance ideologies for bringing about social change via economic-political-religious deviancy – even terrorism. U.S. mainstream media has contributed significantly to the normalization of the 'gangster lifestyle' amongst youths globally [Frazier, 1957; Surette, 1993; Bayles, 1994; Donalson, 2007; Bayles, 2014; Greene, 2016]. The southern U.S. variant of gangster Hip Hop, trap Hip Hop, was marketed via local mainstream media to my peers and me as youths. Many of us embraced the 'trap lifestyle' for neighborhood fame. Yet some of us embraced it for survival amid deepening financial austerity. For such lifestyles are remarkably well-suited to the myriad of jobs warranting long-term subjugation to the U.S. post-chattel slavery, prison-industrial complex (*Shadow Economics + Deep Politics*). These low-level, organized crime jobs – characterized by drug trafficking, sex trafficking, and small-arms trafficking –

MUSIC

perpetuate human sacrifice in the form of continual, gang-related murders [Sarig, 2007; Webb, 2011; Valentine, 2016; Smiley, 2017; Alexander, 2020].

Deeper Levels

Another music form with roots in the southern U.S., Gospel rap, utilizes spoken song and poetic storytelling to promote the message of 'DIVINE Salvation via Jesus THE CHRIST's resurrection' [Lewis, 2009; Ingram, 2016]. Yet a growing number of U.S. clergies, desperate to attract and retain young patrons, are embracing a more spiritually-inclusive substitute [World Council of Churches, 2005; Opsahl, 2016]. This 'holy Hip Hop' puts a Christianized face on conscious Hip Hop. As a chief expression of the Hip Hop movement's light side, conscious Hip Hop utilizes nonverbal and verbal words to promote the message of 'self-salvation via Illumination' (*Sacred Arts & Sciences*). For the eastern U.S. originators of conscious Hip Hop were heavily influenced by Afrocentric innovations upon the inherent Eurocentrism of New Age. New Age has long sought social change via religious-political-scientific activism – even cultism. Alternative media largely based in the West are greatly contributing to conscious Hip Hop's transformation from a regional – to a global – crusade; albeit, in a more ethnically-inclusive form [Black Dot, 2005; Miyakawa, 2005; Lewis and Kemp, 2007; Krs-One, 2009; Atwell, 2012].

Conscious Hip Hop in particular comprised my introduction to Hip Hop in general. Hip Hop deeply resonated with me because its time of birth and ethnic roots are so similar to mine. As a youth, I only knew of Hip Hop as lively

39

dances, catchy rhymes, hypnotic beats, and clever graffiti. Years later, I learned of Hip Hop's role as a civil-resistance movement. More years would pass before I learned of said movement's role as a civil religion. *Indeed, the global Hip Hop movement* – in keeping with civil religions throughout world history – *provides ruling Mystery parties with a powerful means of maximizing mass cooperation with (while simultaneously minimizing mass opposition against) their utopian agendas.*[xvii]

Enter Hip Hop's four 'fundamental elements': writing, instrumentation, dancing, and singing. Uninitiated devotees to the global Hip Hop movement and other non-initiates largely focus on the influential human parties who work with said elements (like entertainment moguls and famous musicians). Yet Hip Hop Initiates and other Mystery Initiates focus on the influential suprahuman parties who empower said elements [Black Dot, 2005; Smith, 2006; Krs-One, 2009; Lewis, 2009; Opsahl, 2016]. For the efficacy of any working involving some empowered element ultimately depends upon the degree to which the worker opens up his or her life to the elemental signified by that element [Rudhyar, 1982; Nicholson, 1987; Thornton, 1995; Ball, 2004; Steiner, 2009].

Singing: In keeping with diverse societies worldwide, those indigenous to Africa utilized repetitive, rhythmic speech in a range of situations; from leisure and labor, to romance and worship. This sacred-profane dynamic persists to today. Some rappers use minstrelized portrayals of life in the world system to entertain listeners with foolishness. Other rappers employ lyrical acrobatics (peppered with embedded marketing as well as inter/intrapersonal critique) to

---

[xvii] The key contrast I make between civil religion and Mystery Religion is that inclusion in the latter is limited to Initiates but inclusion in the former is open to the uninitiated. The key parallel I make is that both speak to the religious-economic-political needs/wants of devotees in such a way that defends the world system' interests.

MUSIC

edutain listeners [Toop, 1984; Walter and Fridman, 2004; Alim, 2006; Stone, 2010]. Yet both groups contain emcees who – as modern enchanters [xviii] – use divinely-inspired speech to administer verbal drugs to listeners and facilitate other incantations. Such sacred speech has been a prominent feature of the Mysteries throughout world history [Rudhyar, 1982; Lewis, 1992; Shaw, 2003; Black Dot, 2005; Filan and Kaldera, 2009; Krs-One, 2009; Sales, 2016].

Dancing: Repetitive, rhythmic, heavenly (as well as earthly) patterns were the bases for ancient African ring-dance. The ritual ring was revered as a space where the sacred is beckoned in – and the profane is expelled out. These rituals persist in a range of forms today; from martial arts like South American capoeira, to healing arts like North American ring shout [Ellfeldt, 1976; Lewis, 1992; Thornton, 1995; Rosenbaum et al., 2013]. Enter the Hip Hop 'cipher': a circular space where psycho-physiological creativity (via dance or song) takes place. Hip Hop dance consists of breaking, waack, etcetera. Most Hip Hop dancers focus on embodying the ups and downs of life in the world system, usually to percussive accompaniment [Alim, 2006; Sarig, 2007; Schloss, 2009; Todd, 2009; George-Graves, 2015]. Yet some Hip Hop dancers focus on embodying the gods. For those divinely-inspired dancers, the cipher is a modern ritual ring where extraordinary psycho-physiological feats take place – often involuntarily. The world system's spirituality is replete with accounts of the gods manifesting through possessed devotees' bodies; often to musical accompaniment [Rudhyar, 1982; Rouget, 1985; Shaw, 2003; Black Dot, 2005; Smith, 2006; Filan and Kaldera, 2009; Lewis, 2009; Staemmler, 2009; Till, 2009].

[xviii] 'Enchantment' centers on beckoning suprahuman presence-power – via repetitive, rhythmic speech – for some covert or clandestine purpose in the natural realm [Elkin, 1977; Faraone, 2001; Walter and Fridman, 2004; Bohak, 2008; Uždavinys, 2009].

Instrumentation: African musical instrumentation is characterized by dense configurations of independent but interrelated rhythms, repetition with subtle variations, and strong emphasis on bass frequencies. It centers on the percussion group's core member: the drum [Blades, 1992; Stone, 2010]. Beatmakers, turntablists, and deejays – as modern drummers – seek to create the ambiances most facilitative of fellowship and edutainment; often via musical EITs [Katz, 2012; Smith, 2013; Schloss, 2014]. Yet some of those instrumentalists seek to create the ambiances most facilitative of workings – especially if involving ASCs – via musical instruments already imbued with the gods' presence-power. Such 'enchanted artifacts' have existed throughout world history in a range of forms; from instruments and ornaments, to monuments and writings [Lethaby, 1891; Rouget, 1985; Arnold, 1989; Blades, 1992; Faraone, 2001; Black Dot, 2005; Smith, 2006; St. John, 2006; Staemmler, 2009; Uždavinys, 2009].

Writing: Symbol systems enabling readers to effectually operate within a social network emerged early in Africa's history. Some open social networks used symbol systems in such a way that hid a message within another message. Some secret social networks utilized stenographic systems to transmute stone, wood, and flesh into conduits for suprahuman presence-power. Furthermore, African Mystery Religion held that such 'enchanted stenography' was imparted to humankind by the 'Queen of Heaven' [Bleeker, 1973; Mafundikwa, 2007]. This hidden-visible dynamic persisted into the African Diaspora. The free U.S. antecedents of Hip Hop's ethnic root wrote profusely and established various social networks. Some of the secret (Christian dominated) social networks used stenography to illegally aid and abet in the liberation of enslaved, fellow blacks. The black Initiates of Atlantic (white dominated) Mystery Religion utilized enchanted stenography not only to advance notions of a global Mystery utopia but also to retain – albeit not always successfully – their own agency as free U.S. citizens [Wesley, 1995; Walker, 2010; Tobin and Dobard, 2011; LaRoche, 2013]. Today, graffers inscribe symbols on flesh, concrete, and steel; largely for identification (like membership in some illegal or legal secret social network), civil-resistance (whether anti-establishment or pro-establishment), and artistic purposes [Gastman and Neelon, 2011; Picerno, 2012]. Yet some of them – as modern hieroglyphic priests – inscribe symbols on the same surfaces for psycho-socio-spiritual transmutation purposes. For Mystery Initiates throughout world history have employed enchanted stenography to facilitate alchemical-imaginal workings [Blavatsky, 1897; Reed, 1972; Brann, 1999; Leet, 1999; Toorn et al., 1999; Black Dot, 2005; Uždavinys, 2010; Bucknell and Stuart-Fox, 2013].

Imagination: On the surface, imagination consists of mental imagery that is imaginary/fantastical. On a deeper level, the imagination comprises the most intimate sacred space where humans can engage – as well as be engaged by – the supranatural realm [Corbin, 1964; Franz, 1979; Steiner, 2009]. Hip Hop Initiates of low degree live contemplative lifestyles centering on at least one of Hip Hop's four distinct fundamental elements (*First Philosophy*). In so doing, they eventually come to attain Hip Hop's unitive, fifth fundamental element: Illumination. Hip Hop Initiates of high degree harness this ultimate element when crafting and conveying the message behind the verbal or nonverbal music. Upon embracing such divinely-inspired messages, TAs (whether cognizant or not) open up their own imaginations to the presence-power of the inspirator [Black Dot, 2005; Krs-One, 2009; Lewis, 2009]. Throughout world history, music has played a vital role in a range of imaginal-alchemical workings; from Initiates' applications of suprahuman presence-power onto TAs and acquisition of hidden knowledge about spiritual reality, to their initiation into the

Mysteries and ascension in degrees. Such training of the mind as a means of transmuting the soul is the objective of most martial arts [Scott, 1935; d'Olivet and Godwin, 1987; Tomlinson, 1994; Thornton, 1995; Shaw, 2003; Walter and Fridman, 2004; Brennan, 2008; Priest and Young, 2010; Saunders, 2012].

The world system's army: War-training has indoctrination and leadership aspects. Both figure prominently in the Mysteries' notion of the god-Man. This Illumined warrior-king commands the human proxies of the world system's army in accordance with *Sacred Law*. This ruling Mystery party – operating above all human law and against DIVINE Law – is thus taught and obligated to defend said system's interests from all threats. The main threats are identified as the GOD of Scripture's Army in general and its human members in particular [Pike, 1874; Bailey, 1957; Arnold, 1992; Hubbard, 1993; Monteith, 2009; Dizdar, 2009; Heiser, 2020]. War-training also has intelligence production and battle engagement aspects. The latter factor figures prominently in the world system's army's exploitation of music via enchantments. The aim is to seize the initiative from enemies and transition into offensive engagement; by musically cultivating-reinforcing spiritual heart issues in enemies (even allies) – as well as fostering the musical ambiances most facilitative of attack. The former factor figures prominently in the world system's army's exploitation of nature via divination. The aim here is to portray misleading views of physical and spiritual reality; by injecting lies into enemies' (even allies') interpretations of natural phenomena – lies bolstered by chance or contrived synchronicities between ecological crises and social crises. Indeed, divination and enchantment are two of said army's most potent psycho-spiritual war-fighting technologies (PWTs).

Most interestingly of all, Scripture intimately associates this and other forms of Revolutionary spiritual warfare with the mass oppression and self-

MUSIC

destruction that increasingly characterizes life in an inherently predatory and insecure world system.

One of my favorite cities

Washington, D.C

SPIRITUALITY

O n the surface, the world system's spirituality is evident. Its deeper levels are hidden behind the Mysteries' multiple layers of physical and nonphysical secrecy. Scripture identifies the ultimate objects of worship in said spirituality as an army of wicked suprahumans. This army has long harnessed said spirituality's colossal Revolutionary spiritual warfare utility to defend its interests: keeping possessions/devotees psycho-spiritually enslaved and subverting GOD's human Family. Interestingly, execution of this defensive war adheres to a pattern: influential or ruling human parties in particular and humankind in general – upon succumbing to the **seduction** of our own spiritual heart issues – knowingly **persecute** the weak and the innocent among us. In so doing, we not only serve as said army's expendable proxy warriors but also place ourselves under a DIVINE Curse long decreed against said army's suprahuman commanders. Evidence of this war pattern exists throughout world history. The multi-ethnic 'nation of nations' that culturally contextualizes both Scripture and

THE CHRIST/MASHIACH is a compelling case in point [Genesis 3; Ezekiel 38-39; Psalm 82; Matthew 15; Galatians 3-5; 2 Corinthians 4; Heiser, 2015].

The most significant instance of said war pattern occurred during the 1st century. By then, Israel had largely come to expect that THE MASHIACH would arrive to usher in a Hebraic utopia. Among the many Israelites asserting themselves to be THE MASHIACH was Yeshua of Nazareth. Yet Yeshua was unique in that He neither promoted Hebrew ethnocultural superiority nor challenged Rome's rulership over Jerusalem [Matthew 13-23; Mark 8-14; Luke 12-19; John 4-18; O'Neill, 1995; Prasch, 2008].

Instead, on the surface, He challenged the *humans* who knowingly failed to execute their DIVINELY-Ordained offices. These ruling Israelite parties were supposed to enact social governance. Yet, in their quest for fame and their love of money, these parties not only conflated the human laws of Hebraic tradition with the DIVINE Law of HOLY Tradition (**seduction**) but also rationalized, romanticized, and legitimized **persecution** of the weak and the innocent among them. Yeshua asserted that such spiritual and moral corruption was preventing fellow Israelites from following – or even recognizing – their long prophesied MASHIACH. Hence He violently shut down the powerful exchange bankers' [xix] lucrative commerce at the Temple's outer court; more than once [Isaiah 9, 52-56; Zechariah 6; Mark 3-12; Luke 19-20; John 2, 17; Helyer, 2002].

On a deeper level, Yeshua challenged the *suprahumans* who knowingly failed to execute their DIVINELY-Ordained offices. These gods are supposed to facilitate social governance within their allotted nations. Yet they deny justice to the weak and the innocent (**persecution**) while showing favor to the wicked

---

[xix] Their ethnocultural antecedents used scholarly methods like checksums to preserve the integrity of various writings; like Old Testament Scripture [Himmelfarb, 2007].

SPIRIT

(**seduction**). Scripture had long prophesied that GOD would Destroy their world system. Hence Yeshua not only liberated countless demonized people but also provoked the powerful gods of the nations into trying to prevent Him from ostensibly fulfilling that prophecy; more than once [Psalm 82, 91; Matthew 11; Mark 3-5; Luke 7-8; John 2; Arnold, 1992; Toorn et al., 1999; Heiser, 2020].

Those provocations – especially the one that occurred at Caesarea Paneas – would prove successful. For the same Yeshua who self-identified as the Way, the Truth, and the Life was soon buried under strict Roman guard following a torturous death sentence; having violated neither human law nor DIVINE Law (**persecution**). Yet, within a few weeks, hundreds of Israelites were testifying to Yeshua's corporeal resurrection! The world system's rulers had been out-flanked. For they were unaware that the Commander of GOD's Army had a secret Agenda: restore humankind's original access to corporeal *Immortality* by exploiting corporeal *death*. This **righteous deception** set the destruction of the world system into motion – on every front [Joshua 5; John 10-19; Hebrews 2; 1 Corinthians 2; Colossians 2; Lapid, 1983; Habermas and Licona, 2004]:

- **Netherworld**: the world system's chief ruler lost control of his top weapon of mass destruction: death – especially, the terror death elicits in humankind.
- **Heaven**: said system's rulers were humiliated in front of fellow suprahumans.
- **Earth**: the earthly manifestation of GOD's Kingdom commenced.

A minority of Israelites came to comprise the first Christians. For they asserted that Yeshua had fulfilled the prophetic purpose of Scriptural Judaism in general and the Temple in particular. For the Temple was where a Levite high priest annually offered unblemished beasts to atone for Israel's wickedness and reestablish socio-ecological order in Israel. Yet those lesser blood sacrifices signified the greatest: a DIVINE High Priest (and Ruler) offering a one-time,

HOLY Sacrifice to atone for all nations' wickedness and reestablish global socio-ecological order [Leviticus 1-27; Isaiah 52-53; Daniel 9; Barker, 2003; Barker, 2010]. Christians asserted Yeshua was said *Sacrificer* as well as said *Sacrificee* – and thus THE MASHIACH – so they were increasingly **persecuted** by the progenitors of a Hebraized civil religion tracing back to the Babylonian exile: Talmudic Judaism. Instead of disputing His resurrection, influential and ruling Israelite parties at work within Talmudic Judaism redefined Yeshua as anything but THE MASHIACH. Yeshua's prophecy of DIVINE Judgment came to pass a few decades later, as revolts by Hebrew nationalists provoked the Romans to destroy the Temple and to expel most Israelites from the land. False prophets/teachers went on to **seduce** generations of fellow Israelites into various syncretic practices; like embracing false mashiachs and journeying the pathways of the gods [Dimont, 1962; Lapid, 1983; Scholem, 1990; Helyer, 2002; Aberbach, 2003; Pardo-Kaplan, 2005; Bohak, 2008; Schäfer, 2009; Bowersock, 2013].

Western Religion

Further instances of the war pattern occurred as the Roman Empire increasingly **persecuted** Christians. Many Christians adapted the true Gospel's message to non-Hebraic social contexts while maintaining its integrity – enabling fellow Gentiles to *hear* GOD's Word. This was in keeping with the first Christians – who enabled fellow Israelites to *see* GOD's Word; namely, that GOD's human Family is open to all nations [1 Peter 1-5; Hurtado, 2005; Frend, 2008]. Yet some Christians conflated human traditions with

SPIRIT

HOLY Tradition – enabling CHRIST-redefining, FATHER-less, HOLY SPIRIT-counterfeiting gospels to erode the earthly manifestation of GOD's Kingdom. The most **seductive** among these false gospels variously either conflated THE CHRIST (in particular) and DIVINE Operations (in general) with creatures; conflated DIVINE Operations with DIVINE Personhood; or portrayed DIVINE Personhood as transferable.[xx] By the 5th century, the now Christianized Roman Empire was **persecuting** fellow (albeit, non-Hellenic) Christians in Western Asia and Northern Africa [Arnold, 1996; McGuckin, 2017; Bantu, 2020].

Other instances occurred during the 10th-14th centuries. Ruling Eastern Roman and post-Western Roman parties had already Christianized most of the Animist nations indigenous to Europe. Yet ruling Mystery parties at work within the Church of the West used the aforesaid false gospels [xxi] to innovate upon the HOLY Tradition preserved in the Church of Eastern Rome, Northern and Eastern Africa, as well as Western and Central Asia [Ware, 1995; Gilman and Klimkeit, 1999; Oden, 2011; Bogdan and Snoek, 2014; Ellingsen, 2015]. Such secularizing teachings – bolstered by events such as the legitimization of counterfeit Roman documents – led to the establishment of a Christianized civil religion: Roman Catholicism. Ruling Roman Catholic parties went on to **seduce** countless fellow Gentiles into various syncretic practices; like embracing

---

[xx] 'DIVINE Operation' refers to GOD's uncreated, *immanent* Activity or Outworking of the DIVINE Good – like Love, Judgment, Holiness, Immortality, Grace, and Revelation – which any creature may manifest (as per the DIVINE Will). 'DIVINE Personhood' refers to GOD's non-transferable, *transcendent* Essence [Ware, 1995].

[xxi] Consider the 3rd century, Modalist gospel portraying THE GODHEAD as consisting of one DIVINE Person. Also consider the 4th century, Arian gospel portraying GOD THE FATHER and GOD THE SON as essentially **two**. Then consider the legendary Roman Catholic teaching of 'absolute divine simplicity' [McGuckin, 2017].

false christs via a Caesar-like papacy and journeying the pathways of the gods. Moreover, while executing holy wars against the Islamic Empire and its rival conception of utopia, this Christian Empire **persecuted** Christians (as well as Jews) across Europe and other parts of the Greater Mediterranean [Maxwell-Stuart, 2005; France, 2006; Valla, 2008; Kelley, 2011; Heather, 2014].

More instances of the war pattern occurred during the 14th-18th centuries. In Europe, Protestantism arose out of Roman Catholicism. Latent seeds of civil religion soon bloomed as the former, in un-Scriptural response to the latter's countless atrocities, innovated upon the aforesaid false gospels [xxii] (**seduction**). In the Americas, ruling Protestant and Roman Catholic parties exploited the many points of similarity between European and indigenous Mystery Religion to push rival notions of a global Christian utopia. Said parties tolerated – even mandated – countless atrocities across the Americas and Europe in the process [Hall, 1951; Churchill, 1997; Yates, 2001; Zafirovski, 2009; Pestana, 2011; Fynn-Paul, 2017]. Maligning Scriptural Christianity as a root cause of those atrocities, ruling Mystery parties advanced Atheist and Animist variants of Scientism. Nevertheless – in keeping with the ancient Church – a remnant of Christians in Europe as well as in the Americas persevered in the faith of THE CHRIST; despite increasingly sophisticated (and often racialized) **persecution** [Cooper, 2006; Melanson, 2009; Bogdan and Snoek, 2014; Gerbner, 2018].

Still more instances occurred during the 18th-20th centuries. Ruling

---

[xxii] Consider the 2nd century, Marcionite gospel portraying the GOD of Israel's HOLY Tradition and the GOD of Gentile Christianity as essentially **two**. Said gospel spread far and wide; via an all-authoritative, de-Judaized, pseudo-canon. Also consider the 5th century, Nestorian gospel portraying GOD THE SON and Jesus of Nazareth as essentially **two**. Then consider the legendary Protestant teachings of 'limited atonement' and 'sola scriptura' [Vasilev, 2014; McGuckin, 2017; Torrance, 2017].

Mystery parties at work within academic and theological institutions harnessed Scientistic literature and paintings to inculcate the Mysteries' gospel of 'self-salvation via Illumined human reason' into the collective conscious of the modern West (**seduction**) [Voegelin, 1968; Collins and Collins, 2006]. Greatly contributing to this syncretizing dynamic was Europe's love-hate relationship with two Asian civil religions. Indeed, Islam – which blends the aforesaid false gospels [xxiii] with Arabian Animism – and Talmudic Judaism's Supranaturalism had already fostered centuries of scientific progress in Europe [Zwemer, 1920; Vickers, 1986; Coudert, 1999; Schuchard, 2002; Masood, 2009; Hall, 2010].

Yet Europe's philosemitism would once again segue into antisemitism. Enter the Nazis. The Nazis exploited the racialized, utopian ideologies already existing in Talmudic Judaism, Islam, and the Church of the West. Moreover, they innovated upon Old and New World methods for developing loyal martial and labor/sex slaves; like forced participation in blood workings from young ages (**persecution**) [Herf, 2009; Probst, 2012; Webman, 2012; Kenny, 2015; Lively and Abrams, 2017]. Such practices and teachings played a vital role in their grandiose endeavor to raise secret armies of Nazi sympathizers and Aryan god-Men. The Mysteries greatly facilitated this by providing protected avenues for influential Nazi, Protestant, Roman Catholic, Muslim, Talmudic Jewish, Animist (namely Shinto and Hindu), and Atheist (namely Soviet) parties to transact – before, during, and after World War II [Antelman, 1974; Manning, 1981; Calverhall, 1991; Nass, 1992; Rosenthal, 1997; Goodrick-Clarke, 1998; Nekrich, 1997; Coogan, 1999; Seagrave and Seagrave, 2003; Reynolds, 2004;

[xxiii] Indeed, the message of Islam – in keeping with the Modalist and Nestorian gospels prevalent in Arabia long prior to Islam – redefines THE GODHEAD in general and THE CHRIST in particular [Durie, 2014; McGuckin, 2017] .

Daniele, 2005; Taha, 2005; Dizdar, 2009; Krüger et al., 2015; Emory, 2020].

As the serpent walked, the epicenter of the greatest proliferation of Supra-naturalistic mass media in modern history shifted from Germany onto U.S. soil. U.S. citizens had long been exposed to un-Scriptural, *Eurocentric* redefinitions of THE CHRIST via the worldwide Protestantistic print media of civil religions like Hyper-Calvinism and British-Israelism [Chamberlain, 1911; Webb, 1976; Williams, 1991; Thompson and Minnicino, 1997; Yates, 2001; Zafirovski, 2009]. Young white Christians were now being exposed to anti-Scriptural, *non-ethnocentric* redefinitions of THE CHRIST via the Supranaturalistic mass media of civil religions like Secular Fundamentalism (utopian Atheism) and the Beat movement (New Age). Young black Christians were being exposed to un-Scriptural, *Afrocentric* redefinitions of THE CHRIST via the Supranaturalistic mass media of civil religions like the Black Muslims and the Black Israelites [Tonkinson, 1995; Nance, 2002; Hedges, 2009; Dorman, 2013; McGowan, 2014]. Both youth-groups often embraced those conceptions of the self and the DIVINE (**seduction**) in un-Scriptural response to rising economic disparity and political terrorism (**persecution**). Such agitative-integrative conceptions are still being spread countrywide; often via the Internet [Hendershott, 2020; Case and Deaton, 2020; Lachman, 2018; Darity and Myers, 1998; Patterson, 1952].

Instances of the war pattern are now occurring in the 21st century. Decades earlier – concurrent with the revival of Animism (via New Age) and Atheism (via Secular Fundamentalism) in the West – Pentecostalism arose out of the revival of U.S. Evangelical Protestantism. Evangelicalism has long dominated U.S. Christianity, while Pentecostalism is well-poised to dominate Christianity worldwide. Yet false teachers, false miracle workers, and false prophets at work within both are increasingly innovating upon the aforesaid false gospels [xxiv] to promote various syncretic practices; like journeying the pathways of the gods and embracing false christs via papal-like clergies. For such **seduction** would not be possible apart from toleration – even endorsement – by clergies who lack genuine fortitude, discernment, and love. Nevertheless – in keeping with the ancient Church – a remnant of Christians worldwide is persevering in the faith of THE CHRIST as well as prevailing in the execution of counter-Revolutionary spiritual warfare; despite increasingly sophisticated **persecution** [Dager, 1990; Orlowski, 2010; Kyle, 2011; Hunt and McMahon, 2013; Allen, 2016].

Greco-Roman Religion

A few centuries before the fall of the Western Roman Empire, some human members of GOD's Army had an insightful spiritual battle. A PWT specialist who was also a labor slave had been trailing them throughout a Greco-Roman colony for many days, loudly announcing them as 'servants of THE MOST HIGH GOD, who proclaim to us the Way of Salvation'. Upon grasping that a demon in her was counterfeiting GOD THE HOLY SPIRIT in them, one of them commanded it – by the DIVINE socio-spiritual Authority of Jesus THE CHRIST – to depart from her. The Pythōnian demon (and, by extension, its lucrative spiritual empowerment) departed right away [Acts 16].

[xxiv] Consider the Prosperity gospel's conflation of material prosperity with GOD's Grace. Also consider the Oneness gospel's portrayal of THE GODHEAD as consisting solely of THE CHRIST. Then consider the Inerrancy gospel's portrayal of the GOD of HOLY Tradition and the GOD of Scripture as essentially **two** [Morris, 2008; Kyle, 2011].

The serpent-god Pythōn was intimately associated with PWTs, Diana/Artemis, and Apollo. Apollo was variously identified as the heavenly Sun-god, the earthly nature-god, and the netherworldly Inner Sun. This tripartite entity was renowned for imparting divine revelation via possessed female devotees. Trained in PWTs from young ages, many of them held positions of authority in a range of Greco-Roman social roles; from personal and political counselors (humanitarian), to biochemical and psychological warfare specialists (military) [Fontenrose, 1959; Johnston, 2001; Wilson, 2004; Mayor, 2008; Graf, 2009].

Female PWT specialists also held positions of authority in various Mystery priesthoods. Said priesthoods promoted a range of ASC-inducing practices and substances whereby fellow Greco-Romans could journey along *the pathways* of the gods; from musical and pansexual workings, to opium-laced alcohol and 'moly' [Toorn et al., 1999; Murray and Wilson, 2004; Hubbard, 2013]. Such was the situation for several centuries, until growing numbers of Greco-Roman devotees to *the Way* began persuading their families and their neighbors – via personal testimonies as well as DIVINELY-Empowered miracles – to abandon worship of the gods (especially Mother Earth) [Arnold, 1989; Arnold, 1996].

Enter a primary function of Apollo's feminine aspect, Diana/Artemis: the vengefully protective, Netherworld god(dess) of blood workings and PWTs. Ruling Greco-Roman parties made use of both in agitative attempts to *suppress* Christianity. They endorsed divine revelations promoting political terrorism against Christians. Yet most Christians were undeterred – even emboldened – by the subsequent **persecutions** [Arnold, 1989; Digeser, 2006; Frend, 2008]. So ruling Greco-Roman parties began shifting to integrative attempts to *co-opt* Christianity. They endorsed the Christianization of blood workings – especially those centering on sexual love – and some sabotaged PWTs. This would give rise to Christian Supranaturalism (**seduction**) [Arnold, 1996; Faraone, 2001]. Yet early Church scholarship (its epicenter being in Northern Africa) yielded a range of theological innovations critical for Christians' corporeal and spiritual survival; from the sexually abstinent lifestyle of monasticism, to the Old and New Testament canon comprising Scripture [Oden, 2007; Ellingsen, 2015].

Phoenician Religion

Phoenicia's quest for both corporeal and spiritual survival via divinely-inspired scholarship yielded a range of techno-cultural innovations; from the symbol system later foundational to the Greek alphabet, to advanced shipbuilding and celestial navigation. Of all their innovations, those involving seafaring were most significant. For Phoenicia's legendary seafaring prowess situated it at the epicenter of an enormously lucrative, transnational flow of foodstuffs and raw materials – even addictive ASC-inducing substances (drug trafficking), military intelligence and war materiel (arms trafficking), as well as sex/labor and martial slaves (human trafficking) [Dvornik, 1974; Knapp, 1991; Moore, 1999; Azize, 2005; Tepić et al., 2011; Fynn-Paul, 2017].

Such were the circumstances in which DIVINE Judgment was declared against the legendary Phoenician city-state of Tyria. Centuries before, Tyrian master builders had meticulously implemented GOD's Design for the Temple. Now, the Tyrians were mocking the Temple's destruction by the Babylonians; intimating that the GOD of Israel was an unreliable and illegitimate FATHER. Furthermore, Tyria's ruler was ascribing DIVINITY to himself due to the GOD-like sociocultural capital he was endowed with. Hence the DIVINE Judgment prophesized the eventual overthrow of Tyria as well as the eventual gathering of a remnant of every nation into GOD's human Family [Ezekiel 24-28].

That same DIVINE Judgment also implicated Sidonia: the foundational Phoenician city-state that, for centuries, relentlessly sought to **seduce** Israel into worshipping the Sun-god Baal [Ezekiel 27-28]. In keeping with other nations throughout Western Asia, Phoenician Baal-worship was characterized by blood workings centering on psycho-spiritual peace and material-corporeal prosperity – especially during times of socio-ecological crises. Most Israelite rulers followed suit and instituted such worship among the Israelite masses. A pattern soon emerged: what would often begin as un-Scriptural worship of the heavenly *FATHER* GOD – following behind the occasional reign of an Israelite ruler who remained loyal to HOLY Tradition – quickly segued into blatant worship of Baal's feminine aspect, the celestial *Mother* god(dess) Anat [1 Kings 11-2 Kings 17; Toorn et al., 1999; Markoe, 2000; Azize, 2005].

The chief instance of that pattern occurred when a sullen Israelite ruler married the emasculating daughter of a Sidonian ruler-high priest. During their reign, the *many* false miracle workers and false prophets of Baal-Anat enjoyed mass popularity and official endorsement in Israel's northern kingdom. Yet the *few* prophets and miracle workers of the GOD of Israel endured mass disdain and official censorship for speaking out against the kingdom's intensifying ignorance, arrogance, and rebellion against the Truth. An escalating series of DIVINE Warnings via *ecological* crises soon culminated in a DIVINE Judgment via *social* crisis: the Assyrians overthrew the northern kingdom of Israel, then expelled most of the Israelites from that area [1 Kings 16-2 Kings 10].

SPIRIT

Babylonian Religion

The Babylonians later overthrew the empires (namely, Assyria and Egypt) adjacent to Israel – the southern kingdom of which had based a Hebraized system of Baal-worship at the Temple in Jerusalem. The Temple's outer court was repurposed by non-Israelites and Israelites for pansexual workings and hosted an enchanted artifact. Israelites repurposed the middle court for workings centering on incense, enchanted stenography, and Tammuz. Traditionally restricted to the Levite priests only, the inner court was now repurposed by ruling Israelite parties for workings involving a shining star. Furthermore, throughout Western Asia, worship of Shining Ones had long been regarded most effectual in the context of human (namely, child) sacrifice. The southern kingdom of Israel's growing ignorance, arrogance, and rebellion against the Truth eventually led to DIVINE Judgment: GOD THE HOLY SPIRIT Departed the Temple. Soon afterward, the Babylonians destroyed the Temple, overthrew the southern kingdom in stages, then expelled most of the Israelites from that area [Hosea 4; 2 Kings 23-25; Ezekiel 6-11; Toorn et al., 1999].

As Israel's DIVINELY-Ordained religion, Scriptural Judaism was partly intended to serve as a polemic against the religions of their neighbors. Indeed, the origins of Mystery Religion can be traced back to Western Asia – especially Mesopotamia. Ruling Mystery parties throughout that region erected stone and wooden pillars (enchanted artifacts) to signify sacred spaces. Said spaces hosted pansexual rituals directed by officially-endorsed priesthoods. These priesthoods were often comprised of prepubertal as well as transvestitic, male prostitutes – respectively signifying both the androgynous and hermaphroditic character of the Mother god(dess) [Toorn et al., 1999; Leick, 2003; Pinches, 2009; Heiser, 2015]. Ruling Babylonian parties had long been exploiting the colossal social engineering utility of money in a range of ways: from taxing sacred prostitution and periodically erasing small farmers' debts, to dominating one of the most valuable religious markets (incense) in Western Asia and instituting some of the first exchange banks in world history. Babylonians apotheosized the most influential among their rulers; naming stars after them. For in life, those rulers likened themselves unto the antediluvian god-Men – people famed for being the first progeny of women and the serpent-gods and the first to fill the Earth with oppression [Black et al., 1992; Heger, 2011; Hudson, 2018; Heiser, 2020].

On the surface, worship of Tammuz – a Sun-god and the child-consort of the Queen of Heaven – centered on entreating the Mother Earth for sustainable livelihoods and households (*corporeal* survival). On a deeper level, worship of Tammuz centered on entreating the god(dess) of the Netherworld – vengeful protectress of the source of Illumination – to awaken the Inner Sun and thereby facilitate transmutation of the soul (*spiritual* survival) [Toorn et al., 1999; Lind, 2015]. Furthermore, the Mother god(dess) has charged a high price throughout world history for such spiritual, mental, corporeal, and material empowerment – the blood, shed in the context of sex and sacrifice, of the most precious among humankind: our children (**persecution**) [Paris, 1992; Bey, 1995; Murray and Roscoe, 1997; Kripal, 1998; Engel, 2006; Streetlove, 2012; Hutsebaut, 2017].

Such deep moral and spiritual darkness is addressed in a DIVINE Judgment against a wicked Babylonian ruler. That human is likened unto a Shining One who, in pursuit of self-glorification and self-gratification, wreaks havoc upon the Earth [Isaiah 14]. That suprahuman is identified elsewhere in Scripture as the originator of Revolution – and the first creature to slander the CREATOR as being a liar. This ability to counterfeit as a messenger of spiritual and moral light comprises influential suprahuman (as well as human) parties' main means of **seducing** vulnerable masses into arrogance, ignorance, and rebellion against the Light of the Truth [Genesis 3; Psalm 2; 2 Corinthians 2, 11].

Egyptian Religion

An Illumined warrior-king led a humanity unified against DIVINE Law to settle Babylonia's future namesake: Babel. In this legendary city, they began erecting a legendary sacred space wherein they nearly fulfilled the Great work. This grandiose endeavor was cut short by a DIVINE Judgement whereby GOD Divided them into nations and Allotted them gods to worship. GOD then Commanded one such creature-worshipper to abandon his Mesopotamian nation and its gods. Upon obeying, GOD Ordained him patriarch of a new nation by which to resume His human Family. Six centuries later, GOD Liberated it from slavery to the nation of Egypt; via ten DIVINE Judgments. On the surface, its *human* **persecutors** were targeted. On a deeper level, Egypt's *suprahuman* rulers were targeted [Genesis 11-Exodus 40].

The first DIVINE Judgment – partly counterfeited by the Pharaoh's PWT specialists (**seduction**) – transmuted the Egyptian creature-worshippers' water into blood; killing most of their fish. The fifth killed most of their livestock. The seventh blighted most of their herbs as well as shattered most of their trees [Exodus 7, 9]. For they had erected sacred wooden pillars signifying Egypt's endeavor to fulfill the Great work. They ascribed both the psychoactive and therapeutic properties of herbs [xxv] to Egypt's gods. They worshipped those gods for every aspect of their sustenance; namely, arable farming, livestock farming, and aquafarming. Crucial to each of those is water. They revered water as the lifeblood of the Mother Earth. This mindset informed their participation in a range of blood workings; from sacred surgery and rites centering on menarche as well as circumcision, to holy war and rites involving the sacrifice of beasts as well as humans [Budge, 1934; Clark, 2000; Pinch, 2004; Monaghan, 2011].

The eighth DIVINE Judgment finished destroying the Egyptian creature-worshippers' green flora. The ninth plunged them, as well as Israelite creature-worshippers, into three days of supranatural darkness. The tenth killed the first-born among both their beasts and their children [Exodus 10-12]. For they worshipped Osiris – a god intimately associated with the 'Green Man' entity still revered worldwide as the Mother Earth's child-consort – as one of the four fundamental elements immanent in yet transcending the natural realm. Each Pharaoh (the serpent on his crown signifying his socio-spiritual authority) was worshipped in life as Osiris reincarnate. Moreover, Osiris was the Netherworld aspect, as well as firstborn son, of a Shining One worshipped as the source of a supranatural light: Illumination. This Sun-god's ostensibly Eternal Life-giving authority was chiefly made manifest in the firstborn among Egyptian creature-worshippers' beasts – and especially in the firstborn among their own children [Bleeker, 1973; Clark, 2000; Matthews, 2000; Naydler, 2004; Pinch, 2004; Dunand and Zivie-Coche, 2005; Uždavinys, 2008; Herrstrom, 2017].

Still more interestingly, all the aforesaid DIVINE Judgments targeted the vengeful protectress of the world system's spirituality: the Mother god(dess). In

---

[xxv] Egypt figured prominently in the trafficking of ASC-inducing, herbal substances; like alcohol, cannabis, opium, and cocaine [Balabanova et al., 1992; Hillman, 2014].

SPIRIT

Egypt, the Mother god(dess) was variously identified as Hekat: god(dess) of the Netherworld; the androgynous cow of Hathor: god(dess) of Illumination; the hermaphroditic hippo of Taweret: god(dess) of motherhood; Sekhmet: god(dess) of the healing arts and the martial arts; Isis: Osiris' feminine aspect and god(dess) of the Sacred Sciences; etcetera. In other nations, the Mother god(dess) was variously identified as Diana/Artemis, Anat, etcetera [Budge, 1934; Eliade, 1965; Toorn et al., 1999; Pinch, 2004; Monaghan, 2011]. This tripartite – netherworldly, earthly, heavenly – entity is still worshipped world-wide and still a key driver of the global collective unconscious. For the Mother god(dess) is the Instructress of, as well as the Initiatrix into, the Great work. All possessions/devotees are therefore tasked with helping to erode-preclude the earthly manifestation of the Kingdom of THE FATHER GOD of Scripture [Blavatsky, 1897; Capra, 1983; Hubbard, 1993; Fersen, 2003; Taylor, 2010].

Most interestingly of all, Scripture intimately associates the Mother god-(dess) with a supranatural – yet impermanent – war whose suprahuman initiator is often described in terms of natural metaphors; such as the serpent-like, river/sea-monster of Ezekiel 28-32.

Donald Winslow Williams: my Uncle and namesake

U.S. Army, 101st Air Division; killed in the Vietnam War

MILITARY ART

There is an ancient, ongoing war to avert the total destruction of the world system. A main objective of this war is to cultivate-reinforce among all humankind the conceptions of material, corporeal, mental, and spiritual empowerment most expedient to said system's rulers – by 2025, as per the deadline set in Bailey [1955] (**strategic level**). The Zeitgeist shepherds this war's overall execution along in space-time (**operational level**). Technique in general – and propaganda in particular – mutually serve to manage the ever-changing criteria for achieving this war's objectives (**tactical level**). Wicked deception coordinates and sustains the prototypical PWT (**logistical level**) that is said system's army's only remaining weapon of mass destruction: Revolution.

Interestingly, this supranatural war can be metaphorized in natural terms. Imagine a massive sea of chaos oscillating in synchrony with hidden currents of wicked deception and furious tides of the Zeitgeist – a murky sea upon which merciless winds of technique churn relentless waves of propaganda. This sea's

Revolutionary effects have long functioned to erode the grounds of humankind's desire to recognize, follow, or even comprehend the Light of the Truth.

The driving forces of shoreline erosion on Earth are sea currents + wind-wave events + tidal events. Tidal flow is strongest during parallel Sun-Moon alignments and is weakest during perpendicular Sun-Moon alignments. This rise-fall flow is elliptically patterned in the open sea and is rectangularly patterned near the shoreline [Davis and Dalrymple, 2012]:

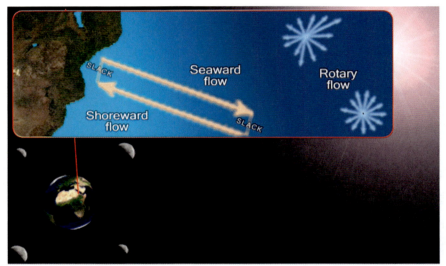

Figure 1. PARALLEL and perpendicular alignments of the Sun and Moon over a lunar month. The inset is of elliptical and rectangular tidal flows over twelve hours.

The fundamental elements of Revolutionary worldview warfare are wicked deception + propaganda-technique + the Zeitgeist. I liken the Zeitgeist to a word-image legendary in the Arts as well as the Sciences:

Figure 2. Diverse societies throughout world history have signified the natural realm as a square (a special case of the rectangle) – and the supranatural realm as a circle (a special case of the ellipse). Similarly, the **collective unconscious** pertains to the world system's perennial-ubiquitous aspects

and the collective conscious pertains to said system's spatial-temporal aspects. Enter the reversible word-image symbolizing humankind's ancient, ongoing pursuit of both self-salvation and global utopia: 'squaring the circle'. 'Circling the square' using only a compass and straightedge – a divinely-inspired method – involves calculating the exact decimal value of the square root ($\sqrt{}$) of the ratio of the circumference of any circle to its diameter (π). Yet obtaining this value of $\sqrt{\pi}$ is mathematically *impossible* [Munari, 2016; Friberg, 2005; Weisstein, 2002; Wilmshurst, 1980; Bruteau, 1974; Seidenberg, 1961].

The Sun drives the main means of redistributing thermal energy across the Earth: wind. Of all wind zones, the trade winds have had the greatest effect on human interaction and interconnectivity. They circulate in a helical pattern:[xxvi]
Figure 3. Helices can be formed with the Euler Identity. It describes two distinct generators of circular motion – via five of mathematics' most significant entities:

- π: the transcendental number governing both the symmetry and the closure of circles.
- **i**: the imaginary unit of complex numbers.
- **0**: the additive identity of real numbers.
- **1**: the multiplicative identity of real numbers.
- **e**: the transcendental number governing the processes of both growth and decay in nature.

The perpendicular projection of a helix onto the plane parallel to its axis is a sine curve. The parallel projection of a helix onto the plane perpendicular to its axis is a prolate wheel curve [Weisstein, 2002; Apostol, 2007; Cockell, 2008].

The Zeitgeist informs both the composition and trajectory of technique. I liken technique to a helix [Ellul, 1980; Ellul, 1986; Demeulemeester, 1994]:
Figure 4. Technique represents an ancient, ongoing process of striving for corporeal and spiritual survival – via five of humankind's most significant and relied-upon works: Divinely-inspired, humanmade innovation (1) that results in the encoding of common

---

[xxvi] The globe with wind cells shown in **Figure 3** is derived from Kaidor [2013].

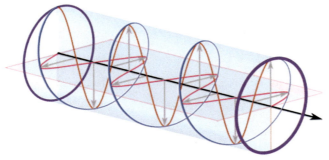

or Mystery culture (2-3) and the production of physical or nonphysical technologies (4-5). Yet **technique** ultimately only tends towards either **secularization** (where we replace DIVINE authority with self-autonomy) or **syncretization** (where we redefine confusion as peace). For we are subject to a de-humanizing – even anti-human – world system and are vulnerable to an *insatiable* desire for self-gratification/self-glorification.

Wind generates shoreline erosion's most conspicuous driving force: waves. As wind energy traverses the surface of the deep sea, water molecules vibrate in a wheel curve [xxvii] pattern [Bascom, 1964; Weisstein, 2002]:

Figure 5. A wheel curve comprises a rotational solution of the Euler Equations for periodic surface gravity water waves. In such a wave, water molecule vibration is both perpendicular and parallel to the wave's direction.

The rhetorical, as well as the most provocative, component of technique is propaganda. I liken propaganda to a prolate wheel curve:

Figure 6. Myths and conditioned reflexes – being constantly revived, repackaged, and refined as needed – comprise sub-**propaganda**. It subtly serves to prepare the TA to actualize

---

[xxvii] Instances of wheel curves also manifest within the drift trajectories of ions in certain electromagnetic fields. Electromagnetic fields are a fundamental physical avenue for mass communication; from antiquity (as in colors) to today (as in radio).

active propaganda some future point. Such actualization may manifest as mass strife (**agitation**); cultivated-reinforced until the TA's breaking point (signified by the top of each narrow loop) is reached. Otherwise, actualization manifests as mass solidarity (**integration**); cultivated-reinforced until a genuine or contrived point of opportunity (signified by the bottom of each wide loop) to **agitate** the TA arises. Yet it is ultimately *impossible* to completely control the human mind via **propaganda** [Ellul, 1965].

Regulating the Earth's climate is a constantly back-forth flowing network of extremely large-scale, cold and warm currents [Garrison and Ellis, 2016]:

Figure 7. Currents in seas' photic zones (from surface to approximately 330 feet) flow counter to currents in seas' aphotic zones. Both are sensitive to localized oscillations in water salinity and temperature. Thus polar waters (unfrozen due to high salinity) flow along the seafloor; below low salinity (due to fresh rain), equatorial waters.

Empowering propaganda is the profoundly emotive, contradictory, and insidious communicative act of wicked deception. I liken wicked deception to an ideal square wave:

Figure 8. In ideal square waves, oscillations between the two levels occur instantly. Yet such oscillation times are *impossible* in physical reality [Weisstein, 2002]. In **wicked deception,** the Lie is seamlessly woven into the Truth.

Scientifically speaking, Earth's biosphere is a union of biological molecules and water molecules. Water molecules can be bonded into diverse compounds; some of which, under certain conditions, exhibit a Borromean-like structure:

Figure 9. This metal-organic weave structure is a humanmade innovation upon the natural geometric-ordering of hydrogen bonds in water clusters. For, in nature,

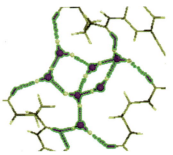
hydrogen bonds tend to self-arrange pyramidically around water molecules. In both of the aforesaid cases, hydrogen bonds serve not only to pull water molecules apart but also to push water molecules together. This vibrational, geometrical dynamic plays a vital role in many of the unique properties making water perfectly suitable for countless technological and biological processes [Russo et al., 2018].[xxviii]

Alchemically speaking, reality is total union of the self and the DIVINE, in essence. Scripturally speaking, this conception of reality constitutes the Lie that leads all who love it into Eternal Destruction. The purpose of Revolution is to cultivate-reinforce said love. I liken Revolution to the Borromean Ring Illusion:

Figure 10. Interlinking three rings such that the center of each is on the boundary of the other two is *impossible*; in physical reality. Yet in the imaginal realm of mathematics, the intersection of such rings forms a shape that can fully rotate in – while always touching all sides of – a square is the Reuleaux triangle. Its three-dimensional analog, the Reuleaux pyramid (**Figure 11**),[xxix] is the intersection of four spheres – the center of each being on the surface of the other three [Lindstrom and Zetterstrom, 1991; Weisstein, 2002].

The deeper any of us fall in love with the world system's conceptions of mental, corporeal, spiritual, and material empowerment – in attempts to satiate our own desires for self-gratification and self-glorification – the higher the risk of our conscience being seared (and thus food for the gods?). At that point, we

---

[xxviii] The water heptamer shown in **Figure 9** is derived from Byrne et al. [2008].

[xxix] The Reuleaux pyramid has an acorn-like shape. So does the pineal gland, which Strassman's pioneering ASC research [2000] refers to as the 'physical seat of the soul'.

effectively cede our own capacity to comprehend, follow, or even recognize the Way of righteousness. Such total de-moralization is a 'necessary evil' of participating in the Great work. For **Revolution** is specifically designed to destroy the facet of the spiritual heart whereby we – based upon our own knowledge of good versus evil (the foremost core belief of every person's worldview) – exercise self-evaluation: the conscience [1 John 2; Exodus 8-9; Matthew 13-15, 24-25; Romans 1-2; 2 Timothy 4; Ephesians 4].

Erosion contributes significantly to planetary homeostasis by regulating beach sediment, much of which dissolves in the waters of the sea. Two of the most prevalent sedimentary ions in seawater are fundamental to humankind's corporeal survival. These ionic compounds self-crystallize into a cubic lattice:

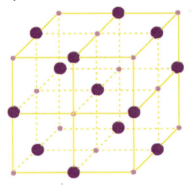

Figure 12. Salt is a cube-shaped, ionically-bonded, sodium-chlorine compound. It has served as a food flavor enhancer and preservative throughout world history [Provost et al., 2016]. In Scripture, salt signifies the preservation of distinctions between the temporal world system and the Eternal Kingdom of GOD; the enhancement of human-to-human and human-to-GOD fellowship; as well as the covenant bond between GOD THE FATHER and GOD's human Family [Leviticus 2; Numbers 18; 2 Chronicles 13; Matthew 5].

Revolution is why the endeavor to fulfill the Great work persists to today. I liken the Great work to my own special case of the Necker Cube Illusion:

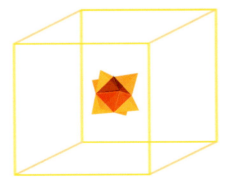

Figure 13. Rhomboid crystals are renowned for eliciting a bistable visual illusion that mathematically corresponds to a rotation through the fourth dimension. A special case of the rhombus, the square, is the second dimension analog of both the cube and its geometric reciprocal: the bipyramid.

Symmetrically extending the vertices and edges of a bipyramid forms an eight-point star: the first step in erecting the both interdimensional and infinite Koch curve.[xxx] In the Mysteries, the eight-point star signifies the Instructress of, as well as Initiatrix into, the **Great work**: the Mother god(dess). The cube signifies a sacred alchemical-imaginal space wherein Ancient Ones impart hidden knowledge of physical and spiritual reality to Initiates. The bipyramid signifies the squared circle/god-Man [Seidenberg, 1961; Leet, 1999; Weisstein, 2002; Pinch, 2004; Friberg, 2005; Pinches, 2009; Wernery, 2013]. Scripturally speaking, the **Great work** and all other false works are like unto psycho-spiritual cages. Inside languish the world system's rulers' slaves: wickedly self-deceived humans. It is *impossible* for them to enter into union with the DIVINE; due to their futile insistence to do so on creatures' terms instead of the CREATOR's [Genesis 3, 11; Isaiah 14, 41-46, 65-66; Mark 8; Romans 1; Galatians 3-5; 1 Corinthians 15].

That **worldview warfare-hydrosphere** metaphor was my innovation upon – as well as my homage to – Ellul's [1965] **propaganda-sea** metaphor. In what follows, I will use his technique characteristics and active propaganda categories to further analyze the war to avert the total destruction of the world system.

The Art of Revolution

The core impetuses behind this prototypical holy war is the prototypical conditioned reflex to DIVINE Law – that it is *suppressing humankind's material, corporeal, mental, and spiritual empowerment* – and the prototypical myth of *god-Men reestablishing Edenic peace and prosperity on their own terms.* Civil religion has long played a vital role in perpetuating this divinely-

---

[xxx] The Koch curve's infinite spikiness suits it well to modeling shorelines; for shorelines naturally exhibit the self-similarity typical of fractals [Mandelbrot, 1977].

MILITARY ART

inspired sub-propaganda in diverse societies worldwide. The global Ecological movement and global Hip Hop movement provide compelling cases in point.

Such highly inclusive, ostensibly leaderless civil religions are quite well suited to horizontal propaganda, in which skilled consensus-builders *subtly steer devotees' 'wrong' ways of thinking and doing into the 'right' ways of thinking* [Ellul, 1965]. Such expedient attitudes and behaviors serve to – as per the universalism of technique [Ellul, 1980] – *intensify devotees' trust and reliance upon technique for managing the insecurities and predation inherent to life in the world system.* The quest for survival has long proved a strong impetus for building consensus. The global Hip Hop movement and global Ecological movement provide devotees with avenues for consensually engaging social and ecological 'evils'. However, they also provide influential human parties with inconspicuous avenues for cultivating-reinforcing expedient conceptions of evil amongst devotees.

On the one hand are the conceptions of evil as a corporeal thing that we must eradicate. Such *worship of death* is – in keeping with rational propaganda [Ellul, 1965] – promoted via a myriad of cold assertions. Indeed, Ecosophy exploits opinions and de-contextualized statistics to rationalize genocidal Eco-legislation as the cure to the 'cancer' that is the weak segment of humankind. On the other hand are the conceptions of evil as an imaginal thing that we must discredit. Such *apotheosization of human reason* is – as per irrational propaganda [Ellul, 1965] – affirmed via a myriad of warm feelings. Indeed, holy Hip Hop exploits emotions and hypnotic ASCs to romanticize Illumination as the key to deciphering the 'fictional' GOD of Scripture. Both of the aforesaid conceptions of evil have long figured prominently in the endeavor to fulfill the Great work. Furthermore, the mere existence of this grandiose endeavor legitimizes – as per the autonomy of technique [Ellul, 1980] – participants' shocking lack of moral compunction for using any means (from thought control to population control) to resolve any threats to its fulfillment.

Ruling human parties often harness genuine and contrived threats to wage worldview warfare among the masses they rule over. Such political propaganda *cultivates-reinforces mass fear and malice in such a way that fosters mass conformity to, and mass reliance upon, ruling human parties' resolutions to said threats* [Ellul, 1965]. The most expedient threat to any collective is that of physical violence; from ecological crises like pandemics, to social crises like terrorism. For the informational, interpersonal, and existential insecurity instigated or exacerbated by real occurrences (even fake news) of such crises provide ruling human parties with key opportunities to advance foreign and domestic political agendas that their societies would otherwise fiercely resist; via legal or even illegal means.

Another way in which the masses *incognizantly unify their own conduct and lifestyles in the manners most expedient to ruling human parties* – a key indicator of sociological propaganda [Ellul, 1965] – is by adhering to conspiracy theories. Socio-behavioral research intimately associates said adherence with psychoses and violence. For said theories divert mass discontent onto scapegoats and exalt fantasies over adherents' lived experiences. As political terrorism and economic disparity intensify globally, mass participation in legal as well as illegal forms of civil resistance is intensifying too. The social engineering-psychological warfare utility of the 'conspiracy theorist' label is sure to grow [Bales, 2007; Blyth, 2013; Bolt, 2011; Webman, 2012].[xxxi] Yet, in the U.S. alone, countless examples exist of ruling human parties harnessing sociological propaganda to advance genuine conspiracies against their own societies [Quigley, 1966; Reisman et al., 1990;

[xxxi] Such elicitation of expedient reactions to – as well as suppression of reasonable and nontrivial analysis of – a person or idea is known as 'name-calling' [Sproule, 2001].

MILITARY ART

Breggin and Breggin, 1998; Iserbyt, 2011; McGowan, 2014; Pepper, 2016; Curry, 2017; Opperman and Redmond, 2017; Collins and Collins, 2020].

Genuine conspiracies and conspiracy theories figure prominently in the Mysteries. Such hierarchical networks are well-suited to vertical propaganda [Ellul, 1965] – for Mystery Initiates of high degree *use secrecy and lies to exploit* Initiates of low degree as well as uninitiated allies. Both of those TAs are – as per the artificiality of technique [Ellul, 1980] – *reduced to pawns through which ruling Mystery parties implement their agendas.* Throughout world history, such lust for power has proved a deeply self-gratifying and self-glorifying impetus for enslaving the other. The Mysteries provide ruling human parties with avenues for imposing their wills upon the masses. However, they also provide the rulers of the world system with inconspicuous avenues for cultivating-reinforcing the most expedient conceptions of the will amongst all humankind.

As Babel shows, there is GOD-like power in a humanity unified in will. Each human's decision to wield said power in a wicked manner is what fuels the world system. That psycho-socio-spiritual dynamic has – as per the monism of technique [Ellul, 1980] – *characterized the application of technē throughout world history.* Furthermore, the resulting technocultural innovations often trace back to the Mysteries. Consider the vital role that the Mysteries have played in so many of world history's most transformative and popular artworks; from songs and monuments, to dramas and texts. Also consider the vital role that the Mysteries have played in so many of world history's most transformative and lucrative professions; namely, the trafficking of drugs, arms, and humans. Then consider how world history's most influential human parties are – as per the rationality of technique [Ellul, 1980] – *impelled* (whether by their own love of money or their own quest for fame) *towards more efficient means to those ends.*

75

REVOLUTION

The *pursuit of progress for progress' sake* – as per the automatism of technique [Ellul, 1980] – pervades the global collective conscious. Scientism has greatly contributed to this psycho-socio-spiritual dynamic. Scientific progress benefits humanity in countless ways. Yet ruling Mystery parties have long exploited those benefits for Revolutionary spiritual warfare purposes. Consider academic-theological institutions' rationalized conflations of pseudoscience with science, alternative-mainstream media's romanticized portrayals of scientific progress as salvific, and government-corporate institutions' legitimized impositions of the Sciences onto areas far beyond scientific purview; like morality or spirituality.

Also consider that the Scientistic gospel and its antecedent, the Supranaturalistic gospel of 'self-salvation via alchemical transmutation', are – as per the inner-impetus of technique [Ellul, 1980] – *driven not by human intervention but rather by the collective unconscious.* On the one hand, such false gospels are deeply appealing to a humanity living under the shadow of an inherently predatory and insecure world system. Indeed, countless people throughout world history have zealously perpetuated false gospels; in pursuit of sociocultural capital, longevity of life, hidden knowledge, an escape from Eternal Destruction, etcetera. On the other hand, the ultimate beneficiaries and originators of all false gospels are wicked suprahumans. Millennia of Supranaturalistic and theological-academic research affirm that those entities do materially, corporeally, mentally, as well as spiritually empower humans. Yet the same research also concedes that this empowerment comes at the expense of those TAs' own wealth, health, familial-social relationships, sanity, even conscience [Arnold, 1989; Shaw, 2003; Collins and Collins, 2020; Gallagher, 2020]. Such *cultivation-reinforcement of moral, intellectual, and affective attitudes facilitative of psycho-spiritual enslavement to the gods of the nations* characterizes what I refer to as 'pneumatological' propaganda.

Then consider the growing anticipation for the emergence of god-Men to establish global utopia [Bailey, 1957; Partridge, 2003; Peterson, 2003; Harari, 2017]. Scripture prophesizes that such an entity will emerge during a time of unprecedented global socio-ecological crises. As the progeny of the Serpent and a woman, this ultimate false mashiach/christ ostensibly fulfills the Great work. Yet this unprecedented global **seduction** rapidly segues into unprecedented global **persecution** once all ruling human parties hand over their authority to him. For another god-Man, the ultimate false prophet, will then mandate all humankind to worship the aforesaid Illumined warrior-king (**DIVINITY**) – or lose the privilege to engage in commerce (**money**) and to even live (**blood**). Yet this global holy war is cut short by the Commander of GOD's Army's descent from Heaven to Jerusalem: the city where the world system's total destruction began. Said system's army, its human proxies having already been gathered there via an unprecedented global enchantment, makes its final effort to subvert GOD's human Family. Still more interesting is how said army's commanders – like the Mother god(dess) and the Serpent who began the Lie – are described post-defeat: as perpetually burnt, *suprahuman sacrifices* unto GOD [Ezekiel 28, 38-39; Psalm 82; Isaiah 14; Apocalypse 13-20; Kline, 1996; Toorn et al., 1999].

The Art of Counter-Revolution

The Lie is foundational to the Great work and all other works that variously locate the ultimate basis (as well as the ultimate means) of Salvation in the creature instead of in THE CREATOR. Chief among these *false works* is the worship of human persons, natural phenomena, supra-human persons, or supranatural phenomena. This creature-worship is carried out in a range of ways; from performing blood workings and awakening the Inner Sun, to engaging in divination and enchantment. Underlying it are a range of *false teachings*; from Scientism and Supranaturalism, to redefinitions of THE CHRIST/MASHIACH and THE GODHEAD. The more a person embraces such teachings and practices, the deeper she or he devolves into spiritual and moral darkness [Genesis 3-11; Leviticus 10; 2 Kings 9; 2 Chronicles 33; Ezekiel 13-21; Jeremiah 5; Romans 1; 2 Corinthians 4-10; Colossians 2; 1 Timothy 4].

THE MOST HIGH GOD – in whom no darkness exists and who cannot lie [Psalm 92; John 8; Titus 1; James 1; Hebrews 6; 1 John 1-2] – communicates the Light of the Truth to humanity in a myriad of ways.[xxxii] Consider the many manifestations of the three DIVINE Persons of THE GODHEAD to the human *senses*. This DIVINE Revelation informed the HOLY Tradition of Israel that centered on the life of the Temple; as preserved until the time of Yeshua/Jesus. This DIVINE Revelation of GOD's Word in human *flesh* informs the HOLY Tradition of GOD's human Family that centers on the life of the Church; as preserved until the time of His future return. Church councils compiled the DIVINE Revelation comprising GOD's Word in human *writing*: Scripture. This DIVINELY-Inspired canon plays a vital role in thoroughly equipping the human members of GOD's Army for counter-Revolutionary worldview warfare. Such

---

[xxxii] Also consider the DIVINE Revelation imparted via DIVINELY-Empowered, physical phenomena like *humanmade* (random outcomes of acts; such as lot-casting), *earthly* (land or sea), and *heavenly* (atmospheric or deep space) events – as well as DIVINELY-Empowered, nonphysical/*imaginal* phenomena like dreams, visions, and promptings. These and all other forms of DIVINE Revelation must operate in accordance with the DIVINE Will [Exodus 3, 16-24; 1 Kings 19; 2 Kings 20; Luke 1, 21; Dam, 1997].

MILITARY ART

DIVINELY-Empowered war-fighting consists of various symbolic-creative human *gestures* and *speech*; like stewarding nature, making music, innovating technologically, liberating the demonized, caring for (even healing) the sick, giving alms to the poor, orderly fellowshipping with fellow Christians, living righteously, praying for people, and sharing the true Gospel [Exodus 33-36; Nehemiah 8; Jeremiah 23-30; Matthew 2-14; Galatians 3-6; 1 Timothy 3; John 1-12; 2 Timothy 3; VanderKam and Flint, 2002; Barker, 2010; Bantu, 2020].

Then consider a special case of DIVINE Revelation: DIVINE Law. Its core component is GOD's Commandments to humankind. They are summarized as: love GOD with all one's soul, physical strength, and spiritual heart; then love fellow humans as much as oneself. All of us are to carry out righteous forms of sexual, familial, and neighborly love towards each other (social governance). Any of us who knowingly neglects the weak – namely the poor, the disabled, the elderly, and born or unborn children – is thus likened unto a wicked suprahuman. Wicked suprahumans, one of the three chief perpetrators of evil, are *the reason why GOD Created a place of Eternal Destruction* (Hell). It follows that their kingdom (the world system) is the second chief perpetrator of evil. For this inherently predatory system centers on cultivating-reinforcing humanity's:

- Avarice, envy of fellow humans' material belongings, and sloth (**money**).
- Malice, lustfulness – especially rape or murder – and unforgiveness (**blood**).
- Ignorance, arrogance, and rebellion against the Truth (**DIVINITY**).

Our own complicity in such wickedness attests that the spiritual heart – having moved away from the DIVINE Good – is a chief perpetrator of evil. Persisting in things having evil effects can so sear our consciences that we cannot express or experience righteous forms of love [Genesis 1-2; Exodus 20, 22; Ezekiel 16; Jeremiah 17; Matthew 15, 22, 25; Mark 7; Hebrews 12; Galatians 5; 1 John 2].

REVOLUTION

The ultimate act of love is to give one's own life for loved ones – whether they be friends or enemies. As the fully-DIVINE, fully-human Jesus/Yeshua of Nazareth willingly suffered the ultimate outcome of wickedness (corporeal as well as spiritual death) having lived a HOLY life, He deprived wickedness of its authority over a Truth-rejecting humanity long subjected to an anti-human world system. For as GOD THE SON returned alive from the Netherworld and ascended (in the flesh) into Heaven, He facilitated the corporeal resurrection of every human. All of us who persevere in His Faith – that is, *wholeheartedly believing in and relying upon DIVINE Salvation* as well as *conscientiously conforming one's own conduct and deeds to the DIVINE Good; despite all the diverse* **persecutions** *and* **seductions** *of life in the world system* – will therefore grow in total union with the DIVINE forever.[xxxiii] **This direct confutation of the Lie comprises the true Gospel.** The true Gospel is implied both in GOD's Covenant with Israel's patriarch and in GOD's Covenant with world history's first man. Most interestingly of all, the true Gospel reflects the prototypical Covenant – made prior to all Creation – among the three Eternally-uncreated, DIVINE Persons of THE GODHEAD: that humankind shall live [Genesis 3, 12-22; Psalm 40, 110; Luke 23; John 5, 15-17; Romans 5, 8; Galatians 3; 1 Corinthians 15; Ephesians 1, 4; 1 Peter 3].

GOD THE HOLY SPIRIT – the Spirit of Truth, who Leads humans into the aforesaid covenant of Eternal Life – is now at work to Persuade all humanity of DIVINE Love and DIVINE Judgment. For although GOD Desires genuine fellowship with each of us, Holiness requires GOD to Judge – even Hate – all

---

[xxxiii] Total union, in this context, refers to humans fully partaking – insofar as it is possible for a creature to do – in GOD's uncreated Operations. For *only GOD THE FATHER, GOD THE SON, and GOD THE HOLY SPIRIT are* **one** *in Essence* [Ware, 1995].

80

MILITARY ART

wickedness. Thus GOD Commands ruling human parties to dispense genuine justice to the weak as well as the powerful. Such social governance includes imposing the death penalty upon murderers of the innocent. Similarly, GOD Commands His human Family to judge the words and deeds of not only false prophets and false miracle workers (without timidity or slander) but also fellow Christians and their own selves (without unforgiveness or whitewashing). GOD THE HOLY SPIRIT is also now at work to Train the human members of GOD's Army. On the one hand, this involves Him Directing the spiritual maturation of untrained warriors – where the rate of their maturation greatly depends on the degree to which they open up their own lives to His Presence-Power. On the other hand, this involves Him Helping trained warriors to stand firm in the sure hope of DIVINE Salvation (defensive) and to boldly proceed in the skillful, prayerful handling of GOD's Word (offensive). Indeed, it is only through such counter-Revolutionary worldview warfare that those HOLY warriors advance the earthly manifestation of GOD's Kingdom and withstand counterattacks from the world system's army [Leviticus 19; Matthew 10; Mark 12; John 3, 14, 16; Romans 6, 8, 13; 1 Corinthians 1-2, 13; 2 Corinthians 1-5, 10; Galatians 5; Ephesians 1-6; Philippians 2; 2 Timothy 2-4; Hebrews 4, 11; James 4].

Those counterattacks' greatest force-multiplier is the TA's own ignorance, arrogance, and rebellion against the Truth. For the originator of these attacks is the father of the Lie in particular and all lies in general. This creature is now at work to cultivate faithlessness among GOD's human Family – by slandering THE FATHER as illegitimate and unreliable. This creature is also now at work to reinforce the psycho-spiritual bondage among people estranged from GOD's human Family – by misrepresenting the self and the DIVINE as essentially **one** [Genesis 3; Psalm 2; Daniel 7; Hebrews 1-2; 2 Timothy 2-3; 1 Peter 5].

My favorite type of place to relax and give thanks

Macao Beach in Punta Cana, Dominican Republic

CONCLUSION

Exploring the Great work is quite humbling. Breaching just its surface layers requires an uncommon psychological education. On the one hand, it features prominently in countless *conspiracy theories*. Such theories drastically frustrate rigorous exploration of it. Also, there is mounting global concern about people who obsess over such theories. For these people increasingly pose a danger to society as well as themselves. On the other hand, the Great work features prominently in the *genuine conspiracy* to avert the total destruction of the world system. Yet the further a person progresses into its deeper levels, the more he or she opens themselves to extreme psycho-spiritual – even corporeal – danger. Hence I strongly caution against exploring it on any level; without the uncommon spiritual force of DIVINE Revelation. Neither do I advise participating in the Great work. For all creatures who endeavor to usurp the CREATOR place themselves under a DIVINE Curse: to become, in Eternal Destruction, ever smaller shadows of what they were originally intended to be.

REVOLUTION

Rather than self-deception, I seek to foster principled, mass awareness that there is a wicked method to – as well as a DIVINE Saviour from – the rapidly escalating madness on Earth. The following suggestions for further exploration build upon the sound foundation that I sincerely hope my book has laid:

- My book explores **spirituality** in the context of *Babylonia, Egypt, Phoenicia, Israel, Greece, Rome,* and the *West*. It can be further explored in the context of Bolivia, Persia, Ethiopia, India, China, Oceania, and the East.

- My book explores **music** via the *philosophical, religious, artistic, political, legal,* and *economic* dimensions of the global Hip Hop movement. It can be further explored via the *scientific* dimension of this movement; like the behavioral and sociological effects of prolonged Hip Hop therapy on fatherless youths.

- My book explores **nature** via the academic, entertainment, theological, and sexual dimensions of the global Ecological movement. It can be further explored via the financial dimension of this movement: carbon pricing.

- My book explores **technological innovation** via the humanitarian-military, mainstream-alternative, and virtual-physical dimensions of EITs. It can be further explored via the corporeal-spiritual dimension: transhumanism.

- My book explores **worldview warfare** on established fronts. It can be further explored on emerging fronts. One such front involves counter-Revolutionary worldview warfare's role in the rapid growth of GOD's human Family among Gentiles in the developing world and among Jews globally. Another involves Revolutionary worldview warfare's role in the rise of UFO-religion globally and the growing legitimization of pansexuality in the developed world.

84

CONCLUSION

My book explores a host of topics about which the half has not been told. I now close with one of the most significant lessons I learned while writing it:

"Remember now your CREATOR in the days of your youth, Before the difficult days come, And the years arrive when you say, 'I have no will to face them'... Remember your CREATOR until the silver cord is removed, And the golden flower is pressed together... Then the dust returns to the earth as it was, And the spirit returns to GOD who gave it... The words of the wise are like pointed sticks, like nails firmly fastened; they are given from the collections and abound from them by one SHEPHERD. Moreover, my son, guard yourself, for there is no end to the making of many books, and much study is weariness of the flesh. Hear the conclusion of the whole matter: Fear GOD and keep His commandment. This is the whole duty of humankind. For GOD will bring every work into judgment, Including everything that has been overlooked, Whether it be good or evil."

[Ecclesiastes 12: 1, 6-7, 11-14]

The birthplace of this book

Jasper, Alabama

BIBLIOGRAPHY

Aberbach D. [2003] The Roman-Jewish Wars and Hebrew Cultural Nationalism, chapter in *Major Turning Points in Jewish Intellectual History*. Palgrave Macmillan, England.

Adeney, F. [1981] Educators Look East, *Spiritual Counterfeits Project Journal*, **4**, [1].

Afropop Worldwide [2016] URL: http://www.afropop.org/ (Accessed: 22nd November, 2016).

Alexander, A. [2020] *The New Jim Crow: Mass Incarceration in the Age of Colorblindness*, The New Press, USA.

Alim, H.S. [2006] *Roc the Mic Right: The Language of Hip Hop Culture*, Routledge, USA.

Allen, J.L. [2016] *The Global War on Christians: Dispatches from the Front Lines of Anti-Christian Persecution*, Penguin Random House, USA.

Andreadaki-Vlazaki, M. (Author); Serra, F. (Editor) [2015] Sacrifices in LM IIIB: early Kydonia Palatial Center, *Pasiphae: magazine of Aegean philology and antiquity*, **IX**.

Antelman, M.S. [1974] *To Eliminate the Opiate*, **1**, Zahavia, USA.

Apostol, T.M (Author); Mnatsakanian, M.A. (Illustrator) [2007] Unwrapping Curves from Cylinders and Cones, chapter in New Horizons in Geometry, *Dolciani Mathematical Expositions*. Mathematical Association of America, USA.

Armytage, W.H.G. [1965] *The Rise of the Technocrats: A Social History*, Routledge, England.

Arnold, C.E. [1989] *Ephesians: Power and Magic: the concept of power in Ephesians in the light of its historical setting*, Cambridge University Press, England.

Arnold, C.E. [1992] *Powers of Darkness: Principalities & Powers in Paul's Letters*, Intervarsity Press, USA.

Arnold, C.E. [1996] *The Colossian Syncretism: The Interface Between Christianity and Folk Belief at Colossae*, Baker Book House, USA.

Ascott, R. [2005] Syncretic Reality: art, process, and potentiality, *Drain Magazine*, **2**, [2].

Atwell, J.M. [2012] *The Transcendentalist Hip Hop Movement*, Master's thesis, Iowa State University, USA.

Austerlitz, P. [1997] *Merengue*, Temple University Press, USA.

Azize, J. [2005] *The Phoenician Solar Theology*, Gorgias Press, USA.

BIBLIOGRAPHY

Bailey, A.A. [1955] *Discipleship in the New Age*, Volume II, Lucis Press, England.

Bailey, A.A. [1957] *The Externalisation of the Hierarchy*, Lucis Publishing Company, USA.

Bainbridge, W.S. [1982] Religions for a Galactic Civilization, chapter in *Science Fiction and Space Futures*. Emme, E.M. (Editor), American Astronautical Society, USA.

Baker, G. [2011] *Buena Vista in the Club: Rap, Reggaetón, and Revolution in Havana*, Duke University Press, USA.

Balabanova, S.; Parche, F.; Pirsig, W. [1992] First identification of drugs in Egyptian mummies, *Naturwissenschaften*, **79**: 358.

Bales, J.M. [2007] Political paranoia v. political realism: on distinguishing between bogus conspiracy theories and genuine conspiratorial politics, *Patterns of Prejudice*, **41**, [1]: 45-60.

Ball, P. [2004] *The Elements: A Very Short Introduction*, Oxford University Press, USA.

Ball, T. [2014] *The Deliberate Corruption of Climate Science*, Stairway Press, USA.

Ballam, M. [1994] *Music and the mind*, Phoenix Productions, USA.

Bantu, V.L. [2020] *A Multitude of All Peoples: Engaging Ancient Christianity's Global Identity*, InterVarsity Press, USA.

Barker, M. [2003] *Great High Priest: The Temple Roots of Christian Liturgy*, T&T Clark, England.

Barker, M. [2010] *Creation: A Biblical Vision for the Environment*, T&T Clark, England.

Bascom, W. [1964] *Waves and beaches: the dynamics of the ocean surface*, Doubleday, USA.

Battalora, J. [2015] *Birth of a White Nation: The Invention of White People and Its Relevance Today*, Strategic Book Publishing, USA.

Bayles, M. [1994] *Hole in Our Soul: The Loss of Beauty and Meaning in American Popular Music*, University of Chicago Press, USA.

Bayles, M. [2014] *Through a Screen Darkly: Popular Culture, Public Diplomacy, and America's Image Abroad*, Yale University Press, USA.

Beebe L.H.; Wyatt T.H. [2009] Guided Imagery and Music: Using the Bonny Method to Evoke Emotion and Access the Unconscious, *Journal of Psychosocial Nursing and Mental Health Services*, **47**, [1].

Beer, W.de [2018] *From Logos to Bios: Evolutionary Theory in Light of Plato, Aristotle & Neoplatonism*, Angelico Press, USA.

Bellin, J.D. [2009] Us or Them!: Silent Spring and the "Big Bug" Films of the 1950s, *Extrapolation: A Journal of Science Fiction and Fantasy*, **50**, [1].

Berg, C. [2017] *Options for addressing instances of ecological harm under the Rome Statute, the added value of an autonomous international crime of ecocide, and its hurdles*, Master's thesis, University of Tilburg, Netherlands.

Bernays, E.L. [1928] *Propaganda*, Horace Liveright Inc., USA.

Berreman, G.D. [1960] Caste in India and the United States, *American Journal of Sociology*, **66**, [2]:120-127.

BIBLIOGRAPHY

Besant, A. [1919] Occultism, *The Theosophist*, **35**, Theosophical Publishing House, India.

Bey, H.; Wilson, P.L. [1995] *Obsessive Love*, Moorish Science Monitor, **7**, [5].

Biehl, J.; Staudenmaier, P. [1995] *Ecofascism: Lessons from the German Experience*, AK Press, USA.

Billington, J.H. [1980] *Fire in the Minds of Men: Origins of the Revolutionary Faith*, Basic Books, USA.

Black Dot [2005] *Hip Hop Decoded*, MOME Publishing Inc., USA.

Black, J.A.; Green, A.; Rickards, T. [1992] *Gods, Demons and Symbols of Ancient Mesopotamia: An Illustrated Dictionary*, University of Texas Press, USA.

Blades, J. [1992] *Percussion Instruments and Their History*, Bold Strummer, USA.

Blainey, G. [1975] *Triumph of the Nomads: A History of Ancient Australia*, Macmillan, Australia.

Blavatsky, H.P. [1897] *The Secret Doctrine: Volume III*, Theosophical Publishing Society, England.

Bleeker, C.J. [1973] *Hathor and Thoth: Two Key Figures of the Ancient Egyptian Religion*, Brill Publishers, Netherlands.

Blyth, M. [2013] *Austerity: The History of a Dangerous Idea*, Oxford University Press, USA.

Bogdan, H.; Snoek, J.A.M. (Editors) [2014] *Brill Handbooks on Contemporary Religion, Vol. 8: Handbook of Freemasonry*, Brill Publishers, Netherlands.

Bohak, G. [2008] *Ancient Jewish Magic*, Cambridge University Press, USA.

Bolt, N. [2011] *Propaganda of the Deed: Are Violent Images a New Strategic Operating Concept in 21st Century Insurgency?*, University of London, England.

Borgeaud, P. (Author); Atlass, K.; Redfield, J. (Translators) [1988] *The Cult of Pan in Ancient Greece*, University of Chicago Press, USA.

Bovermann, T.; de Campo, A.; Egermann, H.; Hardjowirogo, S.; Weinzierl, S. (Editors) [2016] *Musical Instruments in the 21st Century: Identities, Configurations, Practices*, Springer Nature, Singapore.

Bowersock, G.W. [2013] *The Throne of Adulis: Red Sea Wars on the Eve of Islam*, Oxford University Press, USA.

Boyk, J. [2000] *There's Life Above 20 Kilohertz: A Survey of Musical-Instrument Spectra to 102.4 kHz*, California Institute of Technology Music Lab, URL: www.cco.caltech.edu/~boyk/spectra/spectra.htm (Accessed: 5th May, 2020).

Boylan, M. [2015] *The Origins of Ancient Greek Science: Blood - A Philosophical Study*, Routledge, USA.

Brann, N.L. [1999] *Trithemius and Magical Theology: A Chapter in the Controversy over Occult Studies in Early Modern Europe.* State University of New York Press, USA.

Breggin, P.R.; Breggin, G.R. [1998] *The war against children of color: psychiatry targets inner-city youth*, Common Courage Press, USA.

Brennan, T. [2008] *Secular devotion: Afro-Latin music and imperial jazz*, Verso Books, USA.

BIBLIOGRAPHY

Bridenthal, R. [2013] *The Hidden History of Crime, Corruption, and States*, Berghahn Books, USA.

Brooks, J. [2000] Lifestyles: in the Kitchen with Violet, *Daily Mountain Eagle*, August 30.

Brown, C.S. [1970] The Relations between Music and Literature as a Field of Study, *Comparative Literature: Special Number on Music and Literature*, **22**, [2]: 97-107.

Bruteau, B. [1974] *Evolution toward divinity: Teilhard de Chardin and the Hindu traditions*, Theosophical Publishing House, USA.

Brzeziński, Z. [1978] *Between Two Ages: America's Role in the Technotronic Era*, Penguin Books, USA.

Bucknell, R.S.; Stuart-Fox, M. [2013] *The Twilight Language : Explorations in Buddhist Meditation and Symbolism*, Curzon Press, England.

Budge, E.A.W. [1934] *From Fetish to God in Ancient Egypt*, Oxford University Press, England.

Bullock, A. [2004] *The Secret Sales Pitch: An Overview of Subliminal Advertising*, Norwich Publishers, USA.

Byrne, P.; Lloyd, G.O.; Clarke, N.; Steed, J.W. [2008] A "compartmental" Borromean weave coordination polymer exhibiting saturated hydrogen bonding to anions and water cluster inclusion, *Angewandte: Chemie International Edition*, **47**, [31]: 5761-5764.

Cacioppo, J.T; Tassinary, L.G.; Berntson, G.G (Editors) [2007] *Handbook of Psychophysiology*, Cambridge University Press, USA.

Calverhall, R.D. [1991] *Serpent's Walk*, National Alliance, USA.

Capra, F. [1983] *The Turning Point: Science, Society, and the Rising Culture*, Simon and Schuster, USA.

Carson, T.L. [2010] *Lying and Deception: theory and practice*, Oxford University Press, England.

Case, A.; Deaton, A. [2020] *Deaths of Despair and the Future of Capitalism*, Princeton University Press, USA.

Celma, Ò. [2010] *Music Recommendation and Discovery: The Long Tail, Long Fail, and Long Play in the Digital Music Space*, Springer, USA.

Chamberlain, H.S. [1911] *The Foundations of the Nineteenth Century*, J. Lane Publishers, Britain.

Chapoutot, J. [2016] *Greeks, Romans, Germans: How the Nazis Usurped Europe's Classical Past*, University of California Press, USA.

Charet, F. X. [1993] *Spiritualism and the Foundations of C. G. Jung's Psychology*, State University of New York Press, USA.

Chodorow, J. [1991] *Dance therapy and depth psychology*, Routledge, England.

Christensen, T.S. (Editor) [2002] *The Cambridge history of Western music theory*, Cambridge University Press, England.

Churchill, W. [1997] *Little Matter of Genocide: Holocaust and Denial in the Americas 1492 to the Present*, City Lights Books, USA.

Churchill, W. [2004] *Kill the Indian, Save the Man: The Genocidal Impact of American Indian Residential Schools*, City Lights Books, USA.

BIBLIOGRAPHY

Clark, R. [2000] *The sacred tradition in ancient Egypt: the esoteric wisdom revealed*, Llewellyn Worldwide, USA.

Cockell, C. (Editor) [2008] *An Introduction to the Earth-Life System*, Cambridge University Press, England.

Collins, P.D.; Collins, P.D. [2006] *The Ascendancy of the Scientific Dictatorship: An examination of Epistemic Autocracy, From the 19th to the 21st Century*, BookSurge Publishing, USA.

Collins, P.D.; Collins, P.D. [2020] *Invoking the Beyond:: The Kantian Rift, Mythologized Menaces, and the Quest for the New Man*, iUniverse, USA.

Condry, I. [2006] *Hip-hop Japan: Rap and the Paths of Cultural Globalization*, Duke University Press, USA.

Coogan, K. [1999] *Dreamer of the Day: Francis Parker Yockey and the Postwar Fascist International*, Autonomedia, USA.

Cooley, J.W. [2005] Music, Mediation, and Superstrings: The Quest for Universal Harmony, *Journal of Dispute Resolution*, [2]: 227-288.

Cooper, J.W. [2006] *Panentheism – The Other God of the Philosophers: From Plato to the Present*, Baker Academic, USA.

Corbin, H. [1964] Mundus Imaginalis: or the Imaginary and the Imaginal, *Cahiers Internationaux de Symbolisme*, **6**: 3-26.

Coudert, A. (Editor) [1999] *The Impact of the Kabbalah in the Seventeenth Century: the Life and Thought of Francis Mercury van Helmont*, Brill Publishers, Netherlands.

Cramer, P.F. [1998] *Deep Environmental Politics: The Role of Radical*

Environmentalism in Crafting American Environmental Policy, Greenwood Publishing Group, USA.

Crameri, K. [2006] Official, artificial or (arte)factual?: The Museu d'Història de Catalunya, *International Journal of Iberian Studies*, **19**, [2].

Cross, I. [2001], Music, Cognition, Culture, and Evolution, *New York Academy of Sciences: The Biological Foundations of Music*, **930**: 28-42.

Curry, T.J. [2017] *The Man-Not: Race, Class, Genre, and the Dilemmas of Black Manhood*, Temple University Press, USA.

d'Olivet, A.F.; Godwin, J. [1987] *The Secret Lore of Music*, Inner Traditions International, USA.

Dager, A.J. [1990] *Vengeance Is Ours: The Church in Dominion*, Sword Publishers, USA.

Dam, C. [1997] *The Urim and Thummim: A Means of Revelation in Ancient Israel*, Eisenbrauns, USA.

Daniele, G. [2005] *NATO's Secret Armies: Operation GLADIO and Terrorism in Western Europe*, Routledge, USA.

Darity, W.A.; Myers, S.L. [1998] *Persistent Disparity: Race and Economic Inequality in the United States Since 1945*, Edward Elgar Publications, USA.

Darwall, R. [2017] *Green Tyranny: Exposing the Totalitarian Roots of the Climate Industrial Complex*, Encounter Books, USA.

Daskalakis, N.; Georgitseas, P. [2020] *An Introduction to Cryptocurrencies: The Crypto Market Ecosystem*, Routledge, England.

Daube, D. [2011] *Civil Disobedience in Antiquity*, Wipf & Stock Publishers,

BIBLIOGRAPHY

USA.

Davis, D [1986] Ecosophy: The Seduction of Sophia?, *Environmental Ethics*, **8**, [2]: 151-162.

Davis, D.O. [2010] Information Warfare, Globalism, and the Curious Case of Music, *Journal of Information Warfare*, **9**, [1]: 31-45.

Davis, R.A.; Dalrymple, R.W. (Editors) [2012] *Principles of Tidal Sedimentology*, Springer, USA.

Delgado, J.R. [1969] *Physical Control of the Mind: Toward a Psychocivilized Society*, Harper & Row, USA.

DeMeulemeester, J. [1994] *Understanding technique: freedom and determinism in the work of Jacques Ellul*, Master's Thesis, Simon Fraser University, Canada.

Derby, L.H. [2009] *The Dictator's Seduction: Politics and the Popular Imagination in the Era of Trujillo*, Duke University Press, USA.

Dewar, E. [1995] *Cloak of Green: business, government, and the environmental movement*, James Lorimer & Company Ltd., Canada.

Diamond, J. [1997] *Your body doesn't lie: an introduction to behavioural kinesiology*, Eden Grove, USA.

Diethelm, D.; McKee, M. [2009] Denialism: what is it and how should scientists respond?, *The European Journal of Public Health*, **19**, [1]: 2-4.

Digeser, E.D. [2006] Lactantius, Eusebius, and Arnobius: Evidence for the Causes of the Great Persecution, *Studia Patristica*, **39**: 33-46.

Dimont, M.I. [1962] *Jews, God and history*, Signet Book, USA.

Dizdar, R. [2009] *The Black Awakening*, Preemption Books, USA.

Donalson, M.B. [2007] *Hip Hop in American Cinema*, Peter Lang Publishers, USA.

Dorman, J.S. [2013] *Chosen People: The Rise of American Black Israelite Religions*, Oxford University Press, USA.

Dray, P. [2002] *The Hands of Persons Unknown: The Lynching of Black America*, Random House, USA.

Duchesne-Guillemin, M. [1981] Music in Ancient Mesopotamia and Egypt, *World Archaeology*, **12**, [3]: 287-297.

Dunand, F.; Zivie-Coche, C. (Authors); Lorton, D. (Translator) [2005] *Gods and Men in Egypt: 3000 BCE to 395 CE*, Cornell University Press, USA.

Dupré, J. [1993] *The disorder of things: metaphysical foundations of the disunity of science*, Harvard University Press, USA.

Durie, M. [2014] *Which God? Jesus, Holy Spirit, God in Christianity and Islam*, Deror Books, Australia.

Dvornik, F. [1974] *Origins of Intelligence Services: The Ancient Near East, Persia, Greece, Rome, Byzantium, the Arab Muslim Empires, the Mongol Empire, China, Muscovy*, Rutgers University Press, USA.

Dyer, J [2016] *Esoteric Hollywood: Sex, Cults and Symbols in Film*, TrineDay, USA.

Dyer, J [2018] *Esoteric Hollywood II: More Sex, Cults and Symbols in Film*, TrineDay, USA.

Eco, U. [1976] *A Theory of Semiotics*, Indiana University Press, USA.

BIBLIOGRAPHY

Eliade, M. [1962] *The Two and the One*, Havill Press, England.

Eliade, M. [1965] *Mephistopheles and the Androgyne: Studies in Religion Myth and Symbol*, Sheed and Ward, USA.

Elkin, A.P. [1977] *Aboriginal men of high degree: initiation and sorcery in the world's oldest tradition*, University of Queensland Press, Australia.

Ellfeldt, L. [1976] *Dance, from Magic to Art*, W. C. Brown Company, USA.

Ellingsen, M. [2015] *African Christian Mothers and Fathers: Why They Matter for the Church Today*, Wipf and Stock Publishers, USA.

Ellul, J. (Author); Neugroschel, J. (Translator) [1980] *The Technological System*, Continuum Publishing, USA.

Ellul, J. [1965] *Propaganda: The Formation of Men's Attitudes*, A. Knopf, USA.

Ellul, J. [1986] *The Subversion of Christianity*, Wm. B. Eerdmans Publishing, USA.

Emory, D. [2019] *For the Record*, URL: http://spitfirelist.com/category/audio/ (Accessed: 7th May, 2020).

Engel, R. [2006] *The Rite of Sodomy: Homosexuality and the Roman Catholic Church*, New Engel Publishing, USA.

Erdmann, M. [2005] *Building the Kingdom of God on Earth: Churches' Contribution to Marshal Public Support for World Order and Peace, 1919-1945*, Wipf & Stock Publishers, USA.

Erdmann, M. [2009] The Spiritualization of Science, Technology, and Education in a One-World Society, *European Journal of Nanomedicine*, **2**:31-38.

Estabrooks, G.H. [1943] *Hypnotism*, Museum Press, USA.

Faraone, C.A. [2001] *Ancient Greek Love Magic*, Harvard University Press, USA.

Farmer, H.G. [1925] The Influence of Music: from Arabic Sources, *Journal of the Royal Musical Association*, **52**, [1]:89-124.

Feen, R.H. [1996] Keeping the balance: Ancient Greek philosophical concerns with population and environment, *Population and Environment*, **17**:447-458.

Felsen, D.; Kalaitzidis, A. [2005] A Historical Overview of Transnational Crime, chapter in *Handbook of Transnational Crime & Justice*. Reichel, P.; Albanese, J. (Editors), Sage Publications, USA.

Ferguson, M. [1980] *The Aquarian conspiracy: personal and social transformation in the 1980s*, Tarcher Publishing, USA.

Fersen, E. [2003] *Is There A God*, Health Research Books, USA.

Filan, K.; Kaldera, R. [2009] *Drawing Down the Spirits: The Traditions and Techniques of Spirit Possession*, Inner Traditions Publishing, USA.

Finnegan, S. [2013] Eschatological Hedonism: How Asceticism Predisposed Ancient Christians to Reject the Kingdom Hope, article from the *proceedings of the 21st Theological Conference of Atlanta Bible College*, Atlanta, USA.

Fishwick, D. [1992] *The Imperial Cult in the Latin West: Studies in the Ruler Cult of the Western Provinces of the Roman Empire*, Brill Publishers, USA.

Fontenrose, J.E. [1959] *Python: A Study of Delphic Myth and Its Origins*, University of California Press, USA.

Franz, M.L. [1979] *Alchemical Active Imagination*, Spring Publications, USA.

BIBLIOGRAPHY

Frazier, F. [1957] *Black Bourgeoisie*, The Free Press, USA.

Frend, W.H.C. [2008] *Martyrdom and Persecution in the Early Church*, James Clarke and Co Ltd, England.

Friberg, J. [2005] *Unexpected Links Between Egyptian and Babylonian Mathematics*, World Scientific Publishing, Singapore.

Friedman, H.L.; Hartelius, G. (Editors) [2013] *The Wiley-Blackwell Handbook of Transpersonal Psychology*, John Wiley & Sons, England.

Futrell, A. [1997] *Blood in the Arena: The Spectacle of Roman Power*, University of Texas Press, USA.

Fynn-Paul, J. [2017] Empire, Monotheism, and Slavery in the Greater Mediterranean Region from Antiquity to the Early Modern Era, chapter in *Critical Readings on Global Slavery*. Brill Publishers, The Netherlands.

Gallagher, R. [2020] *Demonic Foes: My Twenty-Five Years as a Psychiatrist Investigating Possessions, Diabolic Attacks, and the Paranormal*, Harper-Collins Publishing, USA.

Galton, D.J. [1998] Greek Theories on Eugenics, *Journal of Medical Ethics*, **24**:263-267.

Garrison, T.S.; Ellis, R. [2016] *Essentials of Oceanography*, Cengage Learning, USA.

Gastman, R.; Neelon, C. [2011] *The History of American Graffiti*, Harper-Collins Publishing, USA.

Geissler, H.; Link, S.W.; Townsend, J.T. (Editors) [1992] *Cognition, Information Processing, and Psychophysics: Basic Issues*, Psychology Press, USA.

George-Graves, N. (Editor) [2015] *The Oxford Handbook of Dance and Theater*, Oxford University Press, USA.

Gerbner, K. [2018] *Christian Slavery: Conversion and Race in the Protestant Atlantic World*, University of Pennsylvania Press, USA.

Gesenius, H.W.F.; Tregelles, S.P. [1857] *Gesenius's Hebrew and Chaldee Lexicon to the Old Testament Scriptures*, Samuel Bagster & Sons, England.

Getahun, S.A.; Kassu, W.T. [2014] *Culture and Customs of Ethiopia*, ABC-CLIO, USA.

Gilchrest, E.J. [2013] *Revelation 21-22 in Light of Jewish and Greco-Roman Utopianism*, Brill Publishers, Netherlands.

Gilman, I.; Klimkeit, H.J [1999] *Christians in Asia before 1500*, Routledge, England.

Goldenberg, D.M. [2017] *Black and Slave: The Origins and History of the Curse of Ham*, Walter de Gruyter, Germany.

Goodman, F.D. [1980] Triggering of Altered States of Consciousness as a Group Event: A New Case from Yucatán, *Confmia Psychiatrica*, **23**, [1]: 26-34.

Goodrick-Clarke, N. [1998] *Hitler's Priestess: Savitri Devi, the Hindu-Aryan Myth, and Neo-Nazism*, New York University Press, USA.

Gornitz, V. (Editor) [2009] *Encyclopedia of Paleoclimatology and Ancient Environments*, Springer, Netherlands.

Graf, F. [2009] *Apollo*, Routledge, USA.

Greene, L. [2016] Hip hop producer Afrika Bambaataa molested 'hundreds' of

BIBLIOGRAPHY

kids, bodyguard says: 'There's always a boy in his house', *The New York Daily News*, 6 May, URL: http://www.nydailynews.com/new-york/afrika-bambaataa-molested-hundreds-boys-bodyguard-article-1.2626593 (Accessed: 23rd January, 2020).

Grof, S. [1989] *Spiritual emergency: when personal transformation becomes a crisis*, Penguin Putnam Publishers, USA.

Guénon, R. [2004] *The Reign of Quantity and the Signs of the Times*, Sophia Perennis, USA.

Guliciuc, V.; Guliciuc, E. (Editors) [2010] *Philosophy of Engineering and Artifact in the Digital Age*, Cambridge Scholars Publishing, England.

Habermas, G.R.; Licona, M.R. [2004] *The Case for the Resurrection of Jesus*, Kregel Publications, USA.

Hall, M.P. [1951] *The Adepts in the Western Esoteric Tradition: America's assignment with destiny*, Philosophical Research Society, USA.

Hall, M.P. [2010] *The Adepts in the Eastern Esoteric Tradition: The Mystics of Islam*, Philosophical Research Society, USA.

Hall, T.W.; Edwards K.J. [2002] The Spiritual Assessment Inventory: A Theistic Model and Measure for Assessing Spiritual Development, *Journal for the Scientific Study of Religion*, **41**, [2]: 341-357.

Hanegraaff, W.J. [2018] *New Age Religion and Western Culture: Esotericism in the Mirror of Secular Thought*, Brill Publishers, Netherlands.

Harari, Y.N. [2017] *Homo Deus: A Brief History of Tomorrow*, Harper-Collins Publishing, USA.

Heather, P.J. [2014] *The Restoration of Rome: Barbarian Popes and Imperial Pretenders*, Oxford University Press, England.

Hedges, C. [2009] *When Atheism Becomes Religion: America's New Fundamentalists*, Simon and Schuster, USA.

Heger, P. [2011] *The Development of Incense Cult in Israel*, Walter de Gruyter, Germany.

Heiser, M.S. [2015] *The Unseen Realm: Recovering the Supernatural Worldview of the Bible*, Lexham Press, USA.

Heiser, M.S. [2020] *Demons: What the Bible Really Says about the Powers of Darkness*, Lexham Press, USA.

Helyer, L.R. [2002] *Exploring Jewish Literature of the Second Temple Period: A Guide for New Testament Students*, InterVarsity Press, USA.

Hendershott, A. [2020] *The Politics of Envy*, Sophia Institute Press, USA.

Herf, J. [2009] *Nazi Propaganda for the Arab World*, Yale University Press, USA.

Herlihy, D. [1997] *The Black Death and the Transformation of the West*, Harvard University Press, USA.

Herrstrom, D.S. [2017] *Light as Experience and Imagination from Paleolithic to Roman Times*, Fairleigh Dickinson University Press, USA.

Hillman, D.C.A. [2014] *The Chemical Muse: Drug Use and the Roots of Western Civilization*, Thomas Dunne Books, USA.

Himmelfarb, L. [2007] The Identity of the First Masoretes, *Journal of Sefardic Studies*, **67**, [1]:37-50.

BIBLIOGRAPHY

Hinrichsen, M. [2010] *Jeffersonian Racism*, Ph.D. thesis, University of Hamburg, Germany.

Hollingshead, M. [1974] *The Man who Turned on the World*, Abelard-Schuman, USA.

Hooper, I. [1906] The Reconstruction of Beliefs, from *the proceedings of the Federation of European Sections of the Theosophical Society*, Transactions First Annual Congress, Manen, J. (Editor), Netherlands.

Hope, C. [2008] The possibility of Infrasonic Music, article from the proceedings of the *13th International Conference on Low Frequency Noise and Vibration and its Control*, Tokyo, Japan: 67-79.

Hopkin, B. [1996] *Musical Instrument Design: Practical Information for Instrument Making*, See Sharp Press, USA.

Horne, G. [2015] *Confronting Black Jacobins: The U.S., the Haitian Revolution, and the Origins of the Dominican Republic*, New York University Press, USA.

Horne, G. [2016] *The Counter-Revolution of 1776: Slave Resistance and the Origins of the United States of America*, New York University Press, USA.

Hubbard, B.M. [1993] *The Revelation: Our Crisis Is a Birth*, The Foundation for Conscious Evolution, USA.

Hubbard, T.K. (Editor) [2013] *A Companion to Greek and Roman Sexualities*, John Wiley & Sons, England.

Hudson, M. [2018] *…and forgive them their debts: Lending, Foreclosure and Redemption – From Bronze Age Finance to the Jubilee Year*, Verlag, Germany.

Hughes, J.D. [2014] *Environmental Problems of the Greeks and Romans: Ecology in the Ancient Mediterranean*, John Hopkins University Press, USA.

Hunt, D. (Author); McMahon, T.A. (Editor) [2013] *The Seduction of Christianity*, Berean Call Publishing, USA.

Hurtado, L.W. [2005] *Lord Jesus Christ: Devotion to Jesus in Earliest Christianity*, Wm. B. Eerdmans Publishing, USA.

Husch, J.A. [1984] *Music of the Workplace: A Study of Muzak Culture*, Ph.D. thesis, University of Massachusetts, USA.

Hutchinson, W. [2006] *Information Warfare and Deception*, Informing Science, **9**: 213-223.

Hutsebaut, C. [2017] *Paedofilic networks – victims and offenders – Satanism and rituals*, speech given at the Open Minds Conference, Copenhagen, Denmark.

Ignatiev, N. [2012] *How the Irish Became White*, Routledge, USA.

Immerwahr, D. [2019] *How to Hide an Empire: A History of the Greater United States*, Farrar Straus and Giroux, USA.

Ingram, J. (Director) [2016] *End of Malice*, Ditore Mayo Entertainment, USA.

Isaac, B.H. [2006] *The Invention of Racism in Classical Antiquity*, Princeton University Press, USA.

Isenberg, N. [2016] *White Trash: The 400-Year Untold History of Class in America*, Penguin, USA.

Iserbyt, C.T. [2011] *The Deliberate Dumbing Down of America*, Revised and Abridged Edition, Conscience Press, USA.

BIBLIOGRAPHY

Jacobsen, E.P. [2005] *From Cosmology to Ecology: The Monist World-view in Germany from 1770 to 1930*, Peter Lang Publishers, USA.

Johnson, D.L. [1980] The effects of high level infrasound. Article in the *Proceedings of the Conference on Low Frequency Noise and Hearing*, Moller, H.; Rubak, P. (Editors), Aalborg University Center, Denmark.

Johnston, S.I. [2001] Charming Children: The Use of the Child in Ancient Divination, *Arethusa*, **34**, [1]: 97-117.

Jones, P. [2010] *One or Two: Seeing A World of Difference*, Main Entry, USA.

Jones-Rogers, S.E. [2019] *They Were Her Property: White Women as Slave Owners in the American South*, Yale University Press, USA.

Juda, L. [1978] Negotiating a Treaty on Environmental Modification Warfare: the Convention on Environmental Warfare and its Impact on Arms Control Negotiations, *The International Organization Foundation*, 32, (4):975-991.

Jung, C.G. (Author); Shamdasani, S. (Editor) [2009] *The Red Book*, W.W. Norton & Company, USA.

Juster, S. [2016] *Sacred Violence in Early America*, University of Pennsylvania Press, USA.

Kaidor [2013] *Global circulation of Earth's atmosphere displaying Hadley cell, Ferrell cell and polar cell*, Wikipedia, CC BY-SA 3.0, URL: https://en.wikipedia.org/wiki/File:Earth_Global_Circulation_-_en.svg (Accessed: 15th August, 2019).

Kaku, M. [1999] *Introduction to Superstrings and M-Theory*, Springer, USA.

Karras, R.M. [1988] *Slavery and Society in Medieval Scandinavia*, Yale

University Press, USA.

Katz, M. [2012] *Groove Music: The Art and Culture of the Hip-hop Dj*, Oxford University Press, USA.

Katz, M. [2019] *Build: The Power of Hip Hop Diplomacy in a Divided World*, Oxford University Press, USA.

Kelley, J. [2011] *Anatomyzing Divinity: Studies in Science, Esotericism and Political Theology*, Trine Day, USA.

Kenny, S.C. [2015] Power, opportunism, racism: Human experiments under American slavery, *Endeavour*, **39**, [1]:10-20.

Kimble, G.A.; Boneau, C.A.; Wertheimer, M. (Editors) [1996] *Portraits of Pioneers in Psychology: Volume II*, American Psychological Association, USA.

Kitson, F. [1977] *Bunch of Five*, Faber & Faber Ltd., England.

Kline, M. [1996] Har Magedon: The End of the Millennium, *Journal of the Evangelical Theological Society*, **39**, [2]:207-222.

Knapp, A.B. [1991] Spice, Drugs, Grain, and Grog, chapter in *Bronze Age Trade in the Mediterranean*. Gale, N.H. (Editor), Åströms Förlag, Sweden.

Koestler, A. [1967] *The Ghost in the Machine*, Macmillan, USA.

Koger, L. [2012] *Black Slaveowners: Free Black Slave Masters in South Carolina, 1790-1860*, McFarland Publishers, USA.

Krause, A.E.; North, A.C.; Heritage, B. [2018] Musician interaction via social networking sites: Celebrity attitudes, attachment, and their correlates, *Music & Science*, **1**.

BIBLIOGRAPHY

Krauthamer, B. [2013] *Black Slaves, Indian Masters: Slavery, Emancipation, and Citizenship in the Native American South*, University of North Carolina Press, USA.

Kripal, J.J. [1998] *Kali's Child: The Mystical and the Erotic in the Life and Teachings of Ramakrishna*, University of Chicago Press, USA.

Krippner, S. (Editor) [1978] *Advances in Parapsychological Research: 2 Extrasensory Perception*, Springer, USA.

Krs-One [2009] *The Gospel of Hip Hop*, PowerHouse Books, USA.

Krüger, H. (Author); Meldon, J. (Translator) [2015] *Great Heroin Coup: Drugs, Intelligence & International Fascism*, Trine Day, USA.

Kyle, R. [2011] *Evangelicalism: An Americanized Christianity*, Transaction Publishers, USA.

Lachman, G. [2017] *Lost Knowledge of the Imagination*, Floris Books, England.

Lachman, G. [2018] *Dark Star Rising: Magick and Power in the Age of Trump*, Penguin, USA.

Lampe, K. [2014] *The Birth of Hedonism: The Cyrenaic Philosophers and Pleasure as a Way of Life*, Princeton University Press, USA.

Lapid, P. [1983] *The Resurrection of Jesus: A Jewish Perspective*, Augsburg Fortress Publishers, USA.

LaRoche, C.J. [2013] *Free Black Communities and the Underground Railroad: The Geography of Resistance*, University of Illinois Press, USA.

Lavelle, M. [2016] David Romano Reflects on Mt. Lykaion Discovery, *The American School of Classical Studies at Athens*, URL:

https://www.ascsa.edu.gr/index.php/News/newsDetails/david-romano-reflects-on-mt.-lykaion-discovery (Accessed: 14th November, 2018).

Leet, L. [1999] *The Secret Doctrine of the Kabbalah: Recovering the Key to Hebraic Sacred Science*, Simon and Schuster, USA.

Lehr, J.H. [1992] *Rational Readings on Environmental Concerns*, John Wiley & Sons, USA.

Leick, G. [2003] *Sex and Eroticism in Mesopotamian Literature*, Psychology Press, England.

Leigh, J. [2004] Reflections of Babylon: Intercultural Communication and Globalization in the New World Order, *Globalization*, **4**, [1].

Leprohon, R.J. [2014] Ideology and Propaganda, chapter in *A Companion to Ancient Egyptian Art*. Hartwig, M.K. (Editor), John Wiley & Sons, England.

Lethaby, W.R. [1981] *Architecture, Mysticism and Myth*, Dover Publications, England.

Leventhall, G. [2009] Low Frequency Noise: what we know, what we do not know, and what we would like to know, *Journal of Low Frequency Noise, Vibration and Active Control*, **28**, [2]: 79-104.

Levine, Y. [2018] *Surveillance Valley: The Secret Military History of the Internet*, Hachette Book Group, USA.

Lewis, G.C. [2009] *The Truth Behind Hip-hop*, XulonPress, USA.

Lewis, J.; Kemp, D. (Editors) [2007] *Handbook of New Age*, Brill Publishers, Netherlands.

Lewis, J.L. [1992] *Ring of Liberation: Deceptive Discourse in Brazilian Capoeira*,

BIBLIOGRAPHY

University of Chicago Press, USA.

Liikkanen, L.A. [2008] *Music in Everymind: Commonality of Involuntary Musical Imagery,* Article in the *Proceedings of the Conference on Music Perception and Cognition,* Japan.

Lilly, J.C. [1968] *Programming and Metaprogramming in the Human Biocomputer: Theory and Experiments,* Julian Press, USA.

Lind, R.E. [2015] *The Seat of Consciousness in Ancient Literature,* McFarland Publishers, USA.

Lindstrom, B.; Zetterstrom, H. [1991] Borromean Circles Are Impossible, *The American Mathematical Monthly,* **98**, [4]:340-341.

Lipson, C.S.; Binkely, R.A. [2004] *Rhetoric Before and Beyond the Greeks,* State University of New York Press, USA.

Lively, S.E.; Abrams, K. [2017] *The Pink Swastika: Homosexuality in the Nazi Party,* Veritas Aeterna Press, USA.

Llera, F.J.; Mata, J.M.; Irvin, C.L. [1993] ETA: from Secret Army to Social Movement, the Post-Franco Schism of the Basque Nationalist Movement, *Journal of Terrorism and Political Violence,* **5**, [3]: 106-134.

Lovejoy, P.E. [2011] *Transformations in Slavery: A History of Slavery in Africa,* Cambridge University Press, USA.

Lupyan, G. [2012] What Do Words Do? Toward a Theory of Language-Augmented Thought, chapter in *The Psychology of Learning and Motivation,* Volume 57, Academic Press, USA.

MacArthur, A.P. [2007] The NSA phone call database: The problematic acquisition and mining of call records in the United States, Canada, the United Kingdom, and Australia, *Duke Journal of Comparative and International Law*, **17**, [2]: 441–481.

Mafundikwa, S. [2007] *Afrikan Alphabets: The Story of Writing in Africa*, Mark Batty Publisher, USA.

Mandelbrot, B.B. [1977] *Fractals: Form, Chance, and Dimension*, W. H. Freeman, USA.

Manning, P. [1981] *Martin Bormann: Nazi in Exile*, Lyle Stuart Inc., USA.

Marcos, J. [2005] Just Passing By, *LaGaceta*, 13 July, Salamanca, Spain.

Marin, J.M. [2009] 'Mysticism' in quantum mechanics: the forgotten controversy, *European Journal of Physics*, **30**, [4].

Markoe, G. [2000] *Phoenicians*, University of California Press, USA.

Masood, E. [2009] *Science & Islam: A History*, Icon Books Limited, USA.

Matsumoto, D.R. [2002] *The New Japan: Debunking Seven Cultural Stereotypes*, Intercultural Press, USA.

Matthews, J. [2000] *The Quest for the Green Man*, Theosophical Publishing House, USA.

Maxwell, A. [2010] *Picture Imperfect: Photography and Eugenics, 1870-1940*, Sussex Academic Press, England.

Maxwell-Stuart, P.G. [2005] *The Occult in Medieval Europe 500-1500*, Macmillan Education, England.

Mayor, A. [2008] *Greek Fire, Poison Arrows, and Scorpion Bombs: Biological &*

BIBLIOGRAPHY

Chemical Warfare in the Ancient World, Penguin Books, USA.

McGowan, D. [2014] *Weird Scenes Inside the Canyon: Laurel Canyon, Covert Ops and the Dark Heart of the Hippy Dream*, Headpress Publishers, USA.

McGuckin, J.A. [2017] *The Path of Christianity: The First Thousand Years*, InterVarsity Press, USA.

Meadows, D.H.; Randers, J.; Meadows, D [2004] *Limits to Growth: The 30-Year Update*, Chelsea Green Publishing Company, USA.

Melanson, T. [2009] *Perfectibilists: The 18th Century Bavarian Order of the Illuminati*, Trine Day, USA.

Menezes de Carvalho, E. (Author); Fonseca, L.C. (Translator) [2010] *Semiotics of International Law: Trade and Translation*, Springer, USA.

Meyer, J.; Hansen, U. [2009] *Acoustics and the Performance of Music*, Springer, Germany.

Michaelsen, J. [1982] *The Beautiful Side of Evil*, Harvest House Publishers, USA.

Milton, G. [2012] *White Gold: The Extraordinary Story of Thomas Pellow and North Africa's One Million European Slaves*, Hodder & Stoughton, England.

Minestrelli, C. [2017] *Australian Indigenous Hip Hop: The Politics of Culture, Identity, and Spirituality*, Routledge, England.

Mitchell, T. [2001] *Global Noise: Rap and Hip-hop Outside the USA*, Wesleyan University Press, USA.

Mitchell, T. [2006] Blackfellas Rapping, Breaking and Writing: A Short History of Aboriginal Hip Hop, *Aboriginal History*, **30**: 124-137.

Miyakawa, F.M. [2005] *Five Percenter Rap: God Hop's Music, Message, and Black Muslim Mission*, Indiana University Press, USA.

Molitch-Hou, M. [2019] Paramilitary Panda: WWF Land Grabs Rooted in Covert Apartheid History, *The Reality Institute*, URL: https://therealityinstitute.net/2019/09/paramilitary-panda-wwf-land-grabs-rooted-in-covert-apartheid-history/ (Accessed: 26th November, 2019).

Monaghan, P. (Editor) [2011] *Goddesses in World Culture*, ABC-CLIO, USA.

Monteith, S. [2009] *Brotherhood of Darkness*, Defender Publishing, USA.

Moore, K.; Lewis, D. [1999] *Birth of the multinational: 2000 years of ancient business history*, Copenhagen Business School Press, Denmark.

Morris, J. [2008] *Revival of the Gnostic Heresy: Fundamentalism*, Palgrave Macmillan, USA.

Morris, J.W. [2011] Sounds in the cloud: Cloud computing and the digital music commodity, *First Monday*, **16**, [5].

Mosher, S. [2011] *Population Control: Real Costs, Illusory Benefits*, Transaction Publishers, USA.

Muller, R. [1982] *New Genesis: Shaping a Global Spirituality*, Doubleday, USA.

Munari, B. [2016] *Square, Circle, Triangle*, Princeton Architectural Press, USA.

Murray, P.; Wilson, P. [2004] *Music and the Muses: The Culture of 'mousikē' in the Classical Athenian City*, Oxford University Press, England.

Murray, R.L.; Heumann, J.K. [2009] *Ecology and Popular Film: Cinema on the Edge*, State University of New York Press, USA.

BIBLIOGRAPHY

Murray, S.O.; Roscoe, W. (Editors) [1997] *Islamic Homosexualities: Culture, History, and Literature*, New York University Press, USA.

Music Television Networks [2007] *Diary of Jay-Z in Africa: Water for Life*, URL: http://www.mtv.com/video-clips/xqbuy9/jay-z-supports-water-for-life (Accessed: 15th March, 2018).

Musser, M. [2010] *Nazi Oaks: The Green Sacrificial Offering of the Judeo-Christian Worldview in the Holocaust*, Advantage Inspirational, USA.

Myerson, J.; Petrulionis, S.H.; Walls, L.D. (Editors) [2010] *The Oxford Handbook of Transcendentalism*, Oxford University Press, USA.

Naess, A. [2007] *The Selected Works of Arne Naess: Vols. 1-10*, Springer, USA.

Nance, S. [2002] Mystery of the Moorish Science Temple: Southern Blacks and American Alternative Spirituality in 1920's Chicago, *Religion and American Culture*, **12**, [2]; 123-166.

Nass, M. [1992] Anthrax epizootic in Zimbabwe, 1978–1980: due to deliberate spread?, *Physicians for Social Responsibility Quarterly*, **2**, [4]:198–209.

Naugle, D.K. [2002] *Worldview: The History of a Concept*, Wm. B. Eerdmans Publishing, USA.

Naydler, J. [2004] *Shamanic Wisdom in the Pyramid Texts: The Mystical Tradition of Ancient Egypt*, Simon and Schuster, USA.

Nekrich, A.M. [1997] *Pariahs, Partners, Predators: German-Soviet Relations, 1922-1941*, Columbia University Press, USA.

Nicholson, S.J. [1987] *Shamanism: an expanded view of reality*, Theosophical Publishing House, USA.

Nitzsche, S.A.; Grünzweig, W. [2013] *Hip-Hop in Europe*, LIT Verlag Münster, Germany.

Noll, R. [1994] *The Jung Cult: Origins of a Charismatic Movement*, Princeton University Press, USA.

Noll, R. [1997] *The Aryan Christ: the secret life of Carl Jung*, Random House, USA.

O'Neill, J.C. [1995] *Who Did Jesus Think He Was?*, Brill Publishers, Netherlands.

Oden, T.C. [2007] *How Africa Shaped the Christian Mind: Rediscovering the African Seedbed of Western Christianity*, InterVarsity Press, USA.

Oden, T.C. [2011] *The African Memory of Mark: Reassessing Early Church Tradition*, InterVarsity Press, USA.

Odom, W.E. [2003] *Fixing Intelligence: For a More Secure America - Intelligence in Recent Public Literature*, Yale University Press, USA.

Oohashi, T., Nishina, E.; Honda, M.; Yonekura, Y.; Fuwamoto, Y.; Kawai, N.; Maekawa, T.; Nakamura, S.; Fukuyama, H.; Shibasaki, H. [2000] Inaudible High-Frequency Sounds Affect Brain Activity: Hypersonic Effect, *Journal of Neurophysiology*, **83**, [6]: 3548-3558.

Oohashi, T.; Kawaia, N.; Nishinac, E.; Hondae, M.; Yagia, R.; Nakamuraf, S.; Morimotoe, M.; Maekawah, T.; Yonekurai, Y.; Shibasakik, H. [2006] The role of biological system other than auditory air-conduction in the emergence of the hypersonic effect, *Brain Research*, **1073-1074**: 339-347.

Opperman, E. (Interviewee); Redmond, P. (Interviewer) [2017] Jeffrey Epstein

BIBLIOGRAPHY

Pedophile, *The Opperman Report*, URL: http://www.oppermanreport.com/archive/pearse-redmond-jeffrey-epstein-pedophile (Accessed: 16th April, 2020).

Opsahl, C.P. [2016] *Dance To My Ministry: Exploring Hip-Hop Spirituality*, Vandenhoeck & Ruprecht, Norway.

Orlowski, B.M. [2010] *Spiritual Abuse Recovery: Dynamic Research on Finding a Place of Wholeness*, Wipf and Stock Publishers, USA.

Pardo-Kaplan, D. [2005] Tracing The Antinomian Trajectory Within Sabbatean Messianism, *Kesher: A Journal of Messianic Judaism*, [18].

Paris, G. [1992] *The Sacrament of Abortion*, Spring Publications, USA.

Partridge, C.H. (Editor) [2003] *UFO Religions*, Routledge, England.

Patterson, W.L. (Editor) [1951] *We Charge Genocide: The Crime of Government Against the Negro People*, Civil Rights Congress, USA.

Pedelty, M. [2012] *Ecomusicology: Rock, Folk, and the Environment*, Temple University Press, USA.

Peguero, V. [2004] *The Militarization of Culture in the Dominican Republic: from the Captains General to General Trujillo*, University of Nebraska Press, USA.

Peppard, M. [2011] *The Son of God in the Roman World: Divine Sonship in its Social and Political Context*, Oxford University Press, USA.

Pepper, W.F. [2016] *The Plot to Kill King: The Truth Behind the Assassination of Martin Luther King Jr.*, Simon and Schuster, USA.

Percy, W.A. [1998] *Pederasty and Pedagogy in Archaic Greece*, University of Illinois Press, USA.

Perkins, J. [2004] *Confessions of an Economic Hit Man*, Berrett-Koehler Publishers, USA.

Pestana, C.G. [2011] *Protestant Empire: Religion and the Making of the British Atlantic World*, University of Pennsylvania Press, USA.

Peters, F.E. [1988] Hellenism in Islam, chapter in *Paths from ancient Greece*. Thomas, C.G. (Editor), Brill Publishers, Netherlands.

Peterson, W.S. [2003] *Extraordinary Times, Extraordinary Beings: Experiences of an American Diplomat with Maitreya and the Masters of Wisdom*, Hampton Roads Publishing, USA.

Picerno, D. [2012] *Tattoos: Ancient Traditions, Secret Symbols and Modern Trends*, Chartwell Books, USA.

Pike, A. [1874] *Morals and Dogma*, Masonic Publishing Company, USA.

Pinch, G. [2004] *Egyptian Mythology: A Guide to the Gods, Goddesses, and Traditions of Ancient Egypt*, Oxford University Press, USA.

Pinches, T.G. [2009] *The Religion of Babylonia and Assyria*, BiblioLife, USA.

Pompei, J.F. [2002] *Sound from Ultrasound: The Parametric Array as an Audible Sound Source*, Ph.D. thesis, Massachusetts Institute of Technology in Cambridge, Massachusetts.

Pontifical Academy of Sciences [2016] Evolving Concepts of Nature, proceedings of the *23rd Plenary Sessions of the Pontifical Academy of Sciences*, The Vatican.

Powell, M.A. [2016] *Vanishing America*, Harvard University Press, USA.

Prasch, J.J. [2008] *Israel, the Church and the Jews*, 21st Century Press, USA.

BIBLIOGRAPHY

Priest, G.; Young, D. [2010] *Martial Arts and Philosophy: Beating and Nothingness*, Open Court Publishing, USA.

Probst, C.J. [2012] *Demonizing the Jews: Luther and the Protestant Church in Nazi Germany*, Indiana University Press, USA.

Provost, J.J.; Colabroy, K.L.; Kelly, B.S.; Bodwin, J.; Wallert, M.A. [2016] *The Science of Cooking: Understanding the Biology and Chemistry Behind Food and Cooking*, John Wiley & Sons, USA.

Puett, M.J. [2002] *To Become a God: Cosmology, Sacrifice, and Self-Divinization in Early China*, Harvard University Asia Center, USA.

Quigley, C. [1966] *Tragedy and hope: a history of the world in our time*, Macmillan, USA.

Reed, I. [1972] *Mumbo Jumbo*, Doubleday, USA.

Reisman, J.A.; Eichel, E.W.; Muir, G.J.; Court, J.H. [1990] *Kinsey, Sex, and Fraud: The Indoctrination of a People : an Investigation Into the Human Sexuality Research of Alfred C. Kinsey, Wardell B. Pomeroy, Clyde E. Martin, and Paul H. Gebhard*, Lochinvar-Huntington House, USA.

Reynolds, E. [2004] *Japan in the Fascist Era*, Palgrave Macmillan, England.

Ritenbaugh, J. [2005] *Communication and Leaving Babylon*, Forerunner, USA.

Roederer, J.G. [2008] *The Physics and Psychophysics of Music: An Introduction*, Springer, USA.

Roochnik, D. [2007] *Of Art and Wisdom: Plato's Understanding of Techne*, Penn State Press, USA.

Rosen, C. [2004] *Preaching Eugenics: Religious Leaders and the American*

Eugenics Movement, Oxford University Press, England.

Rosenbaum, A; Rosenbaum, M.N.; Buis, J.S. [2013] *Shout Because You're Free: The African American Ring Shout Tradition in Coastal Georgia*, University of Georgia Press, USA.

Rosenthal, B.G. [1997] *The Occult in Russian and Soviet Culture*, Cornell University Press, USA.

Rosenthal, F. [2015] *Man versus Society in Medieval Islam*, Brill Publishers, Netherlands.

Rouget, G. [1985] *Music and trance: a theory of the relations between music and possession*, University of Chicago Press, USA.

Rudhyar, D. [1982] *The magic of tone and the art of music*, Shambhala Publications, USA.

Russo, J.; Akahane, K.; Tanaka, H. [2018] Water-like anomalies as a function of tetrahedrality, *Proceedings of the National Academy of Sciences of the United States of America*, **115**, [15].

Sachs, C. [1940] *The History of Musical Instruments*, Norton, USA.

Sachs, C. [1943] *The Rise of Music in the Ancient World*, Norton, USA.

Sales, A.D. [2016] The Sources of Authority for Shamanic Speech: Examples from the Kham-Magar of Nepal, *Center for Studies in Oral Tradition: Oral Tradition*, **30**, [2].

Salk, J. [1973] *The Survival of the Wisest*, Harper & Row Publishers, USA.

Sarig, R. [2007] *Third Coast: Outkast, Timbaland, and How Hip-Hop Became a Southern Thing*, Da Capo Press, USA.

BIBLIOGRAPHY

Sathyanarayana, T.S.; Asha, M.R.; Jagannatha, K.S.; Vasudevaraju, P. [2009] The biochemistry of belief, *Indian Journal of Psychiatry*, **51**, [4]: 239-241.

Saunders, D.J. [2012] *Masonic symbols, the ascent to Master Mason, and Wolfgang Amadeus Mozart's 'Maurerische Trauermusik', K. 477*, Master's thesis, University of Alabama, USA.

Schaeffer, F.A. [1976] *How Should We Then Live?: The Rise and Decline of Western Thought and Culture*, Crossway Books, USA.

Schäfer, P. [2009] *Jesus in the Talmud*, Princeton University Press, USA.

Scheuerman, D (2008) From Maxi Voom to Afropop, *Humanities*, **29**, (2).

Schloss, J.G. [2009] *Foundation: B-boys, B-girls and Hip-Hop Culture in New York*, Oxford University Press, USA.

Schloss, J.G. [2014] *Making Beats: the Art of Sample-Based Hip-Hop*, Wesleyan University Press, USA.

Scholem, G. [1990] *Origins of the Kabbalah*, Princeton University Press, USA.

Scholem, G. [1991] *On the Mystical Shape of the Godhead: Basic Concepts in the Kabbalah*, Schocken Books, USA.

Schouhamer, I.K. [2010] Any Song, Anytime, Anywhere, *Journal of the Audio Engineering Society*, **58**.

Schuchard, M.K. [2002] *Restoring the Temple of Vision: Cabalistic Freemasonry and Stuart Culture*, Brill Publishers, Netherlands.

Schuchard, M.K. [2013] *Why Mrs Blake Cried: William Blake and the Erotic Imagination*, Random House, USA.

Schüll, N.D. [2012] *Addiction by Design: Machine Gambling in Las Vegas*, Princeton University Press, USA.

Schuon, F. [1984] *The transcendent unity of religions*, Theosophical Publishing House, USA.

Scott, C. [1935] *An outline of modern occultism*, Routledge, England.

Seagrave, S.; Seagrave, P. [2003] *Gold Warriors: America's Secret Recovery of Yamashita's Gold*, Verso Books, USA.

Seco, L.F.M.; Maspero, G. [2010] *The Brill Dictionary of Gregory of Nyssa*, Brill Publishers, Netherlands.

Seidenberg, A. [1961] The Ritual Origin of Geometry, *Archive for History of Exact Sciences*, **1**: 488-527.

Semetsky, I. [2012] *Jung and Educational Theory*, John Wiley & Sons, USA.

Shaw, G. [2003] Containing Ecstasy: Strategies of Iamblichean Theurgy, *Dionysius*, **22**: 53-88.

Shepherd, G.J.; St. John, J.; Striphas, T. [2006] *Communication as ...: Perspectives on Theory*, Sage Publications, USA.

Sidanius, J.; Peña, Y.; Sawyer, M. [2001] Inclusionary Discrimination: Pigmentocracy and Patriotism in the Dominican Republic, *Political Psychology*, **22**, [4]: 827-851.

Simpson, C. [1994] *The Science of Coercion: Communication Research and Psychological Warfare 1945-1960*, Oxford University Press, England.

Siniossoglou, N. [2011] *Radical Platonism in Byzantium: Illumination and Utopia in Gemistos Plethon*, Cambridge University Press, England.

BIBLIOGRAPHY

Sioshansi, F.P. (Editor) [2011] *Smart Grid: Integrating Renewable, Distributed & Efficient Energy*, Academic Press, USA.

Smiley, C.J. [2017] Addict Rap?: The Shift from Drug Distributor to Drug Consumer in Hip Hop, *Journal of Hip Hop Studies*, **4**, [1]: 94-117.

Smith, K.S. [1995] *Mind, might, and mastery: human potential in metaphysical religion and E. W. Kenyon*, Master's thesis, Liberty University, USA.

Smith, S. [2013] *Hip-Hop Turntablism, Creativity and Collaboration*, Ashgate Publishing, USA.

Smith, W.E. [2006] *Hip Hop as Performance and Ritual*, Trafford, USA.

Snow, N.; Taylor, P.M. [2006] The Revival of the Propaganda State, *International Communication Gazette*, **68**, [5-6]: 389-407.

Sprinkle, A.; Stephens, B. [2017] *The Explorer's Guide to Planet Orgasm: For Every Body*, Greenery Press, USA.

Sproule, J.M. [2001] Authorship and Origins of the Seven Propaganda Devices: A Research Note, *Rhetoric & Public Affairs*, **4**, [1]: 135-143.

St. John, G. [2006] Electronic dance music culture and religion: an overview, *Culture and Religion*, **7**, [1]: 1-25.

St. John, G. [2010] Making a Noise - Making a Difference: Techno-Punk and Terra-ism, *Dancecult: Journal of Electronic Dance Music Culture*, **1**, [2]: 1-28.

Staemmler, B. [2009] *Chinkon Kishin: Mediated Spirit Possession in Japanese New Religions*, LIT Verlag Münster, Germany.

Steiner, R. [2009] *Stages of Higher Knowledge: Imagination, Inspiration, Intuition*, Steiner Books, USA.

Stone, R.M. (Editor) [2010] *Garland Handbook of African Music*, Routledge, USA

Strassman, R. [2000] *DMT: The Spirit Molecule: A Doctor's Revolutionary Research into the Biology of Near-Death and Mystical Experiences*, Simon and Schuster, USA.

Streetlove, C. [2012] *Yesterday's Shame: The Atlanta Child Murders*, Good Ship Publishing, USA.

Surette, L. [1993] *The Birth of Modernism: Ezra Pound, T.S. Eliot, W.B. Yeats, and the Occult*, McGill-Queen University Press, Canada.

Sutton, A.C. [1976] *Wall Street and the Rise of Hitler*, '76 Press, USA.

Svensen, H. [2009] *The end is nigh: a history of natural disasters*, Reaktion Books, England.

Taha, A. [2005] *Nietzsche, Prophet of Nazism: The Cult of the Superman: Unveiling the Nazi Secret Doctrine*, AuthorHouse, USA.

Tate, T. [2019] *Hitler's British Traitors: The Secret History of Spies, Saboteurs and Fifth Columnists*, Icon Books Limited, USA.

Taylor, B.R. [2010] *Dark Green Religion: Nature Spirituality and the Planetary Future*, University of California Press, USA.

Taylor, J. [2007] *Science and omniscience in nineteenth-century literature*, Sussex Academic Press, England.

Tedesko, S. [1992] Family planning media: that's entertainment!, *In Context*, [**31**]:42-43.

Tepić, J.; Tanackov, I.; Stojić, G. [2011] Ancient Logistics: Historical

BIBLIOGRAPHY

Timeline and Etymology, *Tehnicki Vjesnik*, **18**, [3].

Thayer, J.; Strong, J; Grimm, W.; Ludwig, C. [1996] *Thayer's Greek-English Lexicon of the New Testament*, Hendrickson Publishers, USA.

Thomas, T.L.; Elliot, C. [2004] Russian and Chinese Information Warfare: Theory and Practice, *U.S. Army Foreign Military Studies Office*, [A015764].

Thompson, S.; Minnicino, M. [1997] British Israelites and Empire, *Executive Intelligence Review*, **24**, [46]: 36-41.

Thornton, D.A. [1995] *Music in the Mystery Religions of the Ancient World*, University Microfilms International, USA.

Till, R. [2009] Possession Trance Ritual in Electronic Dance Music Culture: A Popular Ritual Technology for Reenchantment, chapter in *Exploring Religion and the Sacred in a Media Age*. Ashgate Publishing Company, USA.

Tilly, M. [1977] The Therapy of Music, chapter in *C.G. Jung Speaking*. McGuire, W.; Hull, R. (Editors), Princeton University Press, USA.

Tobin, J.L.; Dobard, R.G. [2011] *Hidden in Plain View: A Secret Story of Quilts and the Underground Railroad*, Knopf Doubleday Publishing Group, USA.

Todd, M.A. [2009] *Getting krump: reading choreographies of cultural desire through an Afro-diasporic dance*, Ph.D. thesis, Arizona State University, USA.

Tomlinson, G. [1994] *Music in Renaissance Magic: Toward a Historiography of Others*, University of Chicago Press, USA.

Tonkinson, C. [1995] *Big Sky Mind: Buddhism and the Beat Generation*, Riverhead Books, USA.

Toop, D. [1984] *The Rap Attack: African Jive to New York Hip Hop*, Pluto Press

Ltd, USA.

Toorn, K.V.D.; Becking, B.; Horst, P.W.V.D. [1999] *Dictionary of Deities and Demons in the Bible: Second Edition*, Brill Academic Publishers, USA.

Török, L. [2011] *Hellenizing Art in Ancient Nubia 300 B.C. - AD 250 and Its Egyptian Models: A Study in 'Acculturation'*, Brill Publishers, Netherlands.

Torrance, T.F. [2017] *Divine Interpretation: Studies in Medieval and Modern Hermeneutics*, Wipf and Stock Publishers, USA.

Turkle, S. [2005] *Second Self: Computers and the Human Spirit*, MIT Press, USA.

Underhill, E. [1930] *Mysticism: A Study in the Nature and Development of Spiritual Consciousness*, E.P. Dutton & Company, England.

UNESCO [2009] Water in a Changing World, *U.N. World Water Development Report*, [3].

Uždavinys, A. [2004] *The Golden Chain: An Anthology of Pythagorean and Platonic Philosophy*, World Wisdom, USA.

Uždavinys, A. [2008] *Philosophy as a rite of rebirth: from ancient Egypt to neoplatonism*, Prometheus Trust, England.

Uždavinys, A. [2009] Metaphysical symbols and their function in theurgy, *Eye of the Heart Journal*, **2**: 37-59.

Uždavinys, A. [2010] *Philosophy and Theurgy in Late Antiquity*, Sophia Perennis, USA.

Uždavinys, A. [2011] *Ascent to Heaven in Islamic and Jewish Mysticism*, Matheson, England.

BIBLIOGRAPHY

Valentine, D. [2016] *The CIA as Organized Crime: How Illegal Operations Corrupt America and the World*, SCB Distributors, USA.

Valla, L. (Author); Bowersock, G.W. (Translator) [2008] *On the Donation of Constantine*, Harvard University Press, USA.

VanderKam, J.; Flint, P. [2002] *The Meaning of the Dead Sea Scrolls: Their Significance For Understanding the Bible, Judaism, Jesus, and Christianity*, T&T Clark, England

Vasilev, G. [2014] *Heresy and the English Reformation: Bogomil-Cathar Influence on Wycliffe, Langland, Tyndale and Milton*, McFarland Publishers, USA.

Vickers, B. [1986] *Occult Scientific Mentalities*, Cambridge University Press, England.

Virtue, D. [2020] *Deceived No More: How Jesus Led Me out of the New Age and into His Word*, Thomas Nelson Inc, USA.

Voegelin, E. [1968] *Science, Politics and Gnosticism*, Regnery Publishing, USA.

Waal, A. [2017] *Mass Starvation: The History and Future of Famine*, John Wiley & Sons, USA.

Walker, C.D.B. [2010] *A Noble Fight: African American Freemasonry and the Struggle for Democracy in America*, University of Illinois Press, USA.

Wallerstein, I.M. [2004] *World-systems Analysis: An Introduction*, Duke University Press, USA.

Walter, M.; Fridman, E. [2004] Music in World Shamanism. In *Shamanism: An Encyclopedia of World Beliefs, Practices, and Culture*, ABC-CLIO, USA.

Ware, K. [1995] *The Orthodox Way*, Saint Vladimir's Seminary Press, USA.

Webb, G. [2011] *Dark Alliance: The CIA, the Contras, and the Cocaine Explosion*, Seven Stories Press, USA.

Webb, J. [1976] The *Occult Establishment*, Open Court Publishing, USA.

Webman, E. [2012] *The Global Impact of the Protocols of the Elders of Zion: A Century-Old Myth*, Routledge, USA.

Weikart, R. [2006] *From Darwin to Hitler: Evolutionary Ethics, Eugenics, and Racism in Germany*, Palgrave Macmillan, USA.

Weikart, R. [2016] *Hitler's Religion: The Twisted Beliefs that Drove the Third Reich*, Simon and Schuster, USA.

Weindling, P. [1989] *Health, Race and German Politics Between National Unification and Nazism, 1870-1945*, Cambridge University Press, England.

Weisstein, E.W. [2002] *CRC Concise Encyclopedia of Mathematics*, CRC Press, USA.

Welch, K.E. [1987] Keywords from Classical Rhetoric: The Example of Physis, *Rhetoric Society Quarterly*, 17: 193- 204.

Wernery, J. [2013] *Bistable perception of the Necker cube in the context of cognition & personality*, Ph.D. thesis, Swiss Federal Institute of Technology in Zurich, Switzerland.

Wesley, D.P. [1995] *Early Negro Writing 1760-1837*, Black Classic Press, USA.

White, L. [1967] The Historical Roots of Our Ecological Crisis, *Science*, 155:1203-1207.

Whitman, J.Q. [2017] *Hitler's American Model: The United States and the*

BIBLIOGRAPHY

Making of Nazi Race Law, Princeton University Press, USA.

Wierzbicki, A.P. [2015] *Technen: Elements of Recent History of Information Technologies with Epistemological Conclusions*, Springer, Switzerland.

Williams, B. [1991] Apartheid in South Africa: Calvin's Legacy?, *The Upsilonian*: University of the Cumberlands, [3].

Wilmshurst, W.L. [1980] *The Meaning of Masonry*, Gramercy, USA.

Wilson, J.F. [2004] *Caesarea Philippi: Banias, the Lost City of Pan*, I.B. Tauris Publishers, England.

Winkelman, M. [2003] Complementary therapy for addiction: Drumming out Drugs, *American Journal of Public Health*, **93**, [4]: 647-651.

Wolin, S.S. [2009] *Politics and Vision: Continuity and Innovation in Western Political Thought*, Princeton University Press, USA.

Woodard, V. [2014] *The Delectable Negro: Human Consumption and Homoeroticism Within US Slave Culture*, New York University Press, USA.

World Council of Churches [2005] *Ecumenical Courier*, **65**, [1].

Wyatt-Brown, B. [1982] *Southern Honor: Ethics and Behavior in the Old South*, Oxford University Press, USA.

Yates, F.A. [2001] *The occult philosophy in the Elizabethan age*, Routledge, USA.

Zafirovski, M. [2009] *The Destiny of Modern Societies: the Calvinist Predestination of a New Society*, Brill Publishers, Netherlands.

Zwemer, S.M. [1920] *The Influence of Animism on Islam*, Macmillan, USA.

INDEX

A

Altered States of Consciousness
15, 20, 21, 29, 31, 42, 56, 57, 73

Ancient Ones 6, 7, 19, 72

Pan...22

The Green Man 62

Arms Trafficking 23, 38, 75

Intelligence 57

Materiel 57

Art...13

Acting 15, 22, 75

Architecture...15, 22, 42, 59, 61, 75

Fashion 37, 38, 42

Literature 6, 15, 22, 42, 53, 75

Mass Media 15, 18, 35, 54

Painting 15, 22, 53

Stenography 42, 43, 59

Storytelling.......................15, 22, 39

B

Babylonia 50, 57, 59, 60, 61

Babel 15, 32, 61, 75

Biological Warfare...............11, 56

Britain 25, 26

C

Chemical Warfare 56

Civil Religion

   Black Israelites............................ 54

   Black Muslims............................. 54

   Hinduism 53

   Islam.......................... 22, 23, 52, 53

   Nazism 26, 53

   New Age .26, 39, 72, *See* Ecological and Hip Hop Movements

      Beat Movement 54

   Protestantism 52, 53, 54

      British-Israelism 54

      Evangelicalism 54

      Hyper-Calvinism.................... 54

      Pentecostalism 54

   Roman Catholicism ...2, 22, 24, 36, 51, 52, 53

   Roman Imperial Cult 21

   Secular Fundamentalism 54

   Shintoism 53

   Talmudic Judaism 50, 53

Conspiracy

   Conspiracy Theories 5, 6, 74, 83

   Genuine Conspiracies ...83, *See* The Great Work

Culture

   Common Culture .... 13, 68, *See* Philosophy, Religion, Art, Law, Politics, Science, and Economics

   Mystery Culture .....13, 68, *See* First Philosophy, Mystery Religion, Sacred Arts & Sciences, Sacred Law, Deep Politics, and Shadow Economics

D

Death 14

   Corporeal Death........20, 49, 80, 81

   Death-Worship 21, 73

      Human Sacrifice .21, 39, 59, 60, 62, 77

   Spiritual Death/Eternal Destruction 70, 76, 79, 80, 83

      Suprahuman Sacrifice 77

   Terror of Death 49

Deception

   Righteous Deception...................49

   Wicked Deception.................10, 69

      Counterfeit ...11, 17, 34, 51, 55, 61

      Self-Deception.................72, 84

Deep Politics.................13, 26, 38

INDEX

Demonization 32, 41, 56

    Liberating the Demonized 49, 79

Demons......................7, 11, 55, 79

Divination 44, 78

Drug Trafficking..3, 23, 38, 57, 75

    Alcohol 22, 56, 62

    Cannabis..............................22, 62

    Cocaine..................................62

    Moly 56

    Opium 21, 22, 56, 62

E

Ecological Movement...16, 18, 22, 26, 73

    Eco-Edutainment......18, 22, 27, 35

    Eco-Legislation 26, 27, 73

    Eco-Pansexuality 22

    Ecosophy 18, 21, 73

        World Soul 18

    Environmentalism 21, 24, 26

        Blood and Soil Ideology 26

    Socio-Ecological Justice 36

    Sustainable Development .6, 21, 26, 35, 36, 60

Economics 13, 17

    Austerity 37, 38

    Commerce 23, 25, 48, 57, 75, 77

Consumerism...............................38

Disparity 23, 37, 54, 74

    Generational Poverty ...3, 21, 34, 37

Egypt 59, 61, 62, 63

Enchantment........... 41, 44, 77, 78

    Artifacts 42, 59

    Symbols............................42, 43, 59

F

First Philosophy............. 13, 24, 43

    Supranaturalism ..19, 24, 26, 31, 53, 54, 57, 76, 78

G

Germany 24, 26, 54

Godhead (The)........51, 53, 55, 78

    Divine Essence

        Divine Personhood/Divinity 11, 20, 51, 57, 77, 78, 79, 80

    Divine Operations 51

        Grace................................55, 72

        Holiness 3, 19, 50, 79, 80

        Immortality 21, 49, 62, 80

        Judgment.50, 57, 58, 59, 60, 61, 62, 80

        Love 80

        Revelation 44, 78, 79, 83

God the Father... 14, 23, 57, 58, 63, 80, 81

God the Holy Spirit .. 55, 59, 80, 81

God the Son ... 3, 48, 49, 52, 55, 78, 80

Great Work (The) 6, 7, 10, 14, 17, 20, 34, 61, 63, 71, 72, 77, 83

god-Men 24, 44, 72, 77

Aryan 53

Babylonian 60

Self-Salvation ..... 23, 67, 76, 77, *See* Illumination

Utopia..... 15, 25, 40, 42, 48, 52, 53, 54, 67, 77

Greece 20, 21, 51, 55, 56

H

Hip Hop Movement..... 34, 35, 73

Cipher 37, 41

Conscious Hip Hop 37, 39

Fifth Fundamental Element 43, *See* Illumination

Four Fundamental Elements 40, 43

Deejaying............................... 42

Graffiti Writing.................... 43

Hip Hop Dance......... 37, 38, 41

Rapping 40

Gangster Hip Hop 37, 38

Hip Hop Initiate 35, 40, 43

holy Hip Hop 39, 73

Trap Hip Hop 38

Human Trafficking........ 23, 38, 75

Laborial and Sexual 2, 3, 26, 34, 38, 42, 53, 55, 57, 61

Martial................................... 53, 57

Psycho-Spiritual .. 32, 47, 72, 76, 81

Slave Networks

Trans-Atlantic 24

Trans-Mediterranean 23, 57

Trans-Pacific 24

Trans-Saharan 24

I

Illumination 21, 39, 44, 53, 61, 62, 63, 73, 77

Awakening the Inner Sun 6, 56, 60, 78

Sun-god..................... 56, 58, 60, 62

Israel.......50, 57, 58, 59, 61, 78, 80

Jerusalem 48, 59, 77

K

Kingdom of God (The) 47, 49, 51, 63, 71, 81

Army........................ 55, 77, 78, 81

INDEX

God's Human Family.....47, 48, 50, 58, 61, 77, 78, 81

Holy Tradition ...14, 32, 48, 51, 52, 55, 78

The Church 52, 55, 78

Celtic 23

Central Asia............................51

Eastern Africa.........................51

Northern Africa 23, 51, 57

The East 51

The West...................24, 51, 53

Western Asia 51

The Temple...48, 49, 50, 57, 59, 78

The True Gospel.......39, 50, 79, 80

The Truth (God the Son) 10, 49, 58, 59, 61, 66, 69, 71, 78, 79, 81

L

Law ..13, 25, 42, 48, 51, 73, 74, 76

Organized Crime.........3, 15, 38, 44

M

Mathematics 4, 67, 71

Cryptography...........................5, 31

Geometry

Bipyramid 71, 72

Circle 66, 72

Cube 71, 72

Eight Point Star.....................72

Ellipse 66

Euler Equations 68

Euler Identity.........................67

Helix 67

Ideal Square Wave 69

Koch Curve 72

Pyramid 70

Rectangle...............................66

Rhombus................................71

Sphere 70

Square 66, 70, 71, 72

Triangle.................................70

Wheel Curve 67

Mind Control 15, 32, 69

Music 4, 22, 79

Dancing 29

Capoeira.............................41

Merengue............................37

Ring Shout 41

Frequency 18, 29, 30, 31, 33, 42

Gesticulation.........................29, 31

Hip Hop .. *See* Hip Hop Movement

Instrumentation ...29, 30, 32, 37, 42

Musical Artifacts.........................29

Musical Imagery 29

Musician 15, 33, 40

135

Positioning 29

Punk Rock 36, 37, 38

Repetition 29, 31, 40, 41, 42

Rhythm ... 29, 30, 31, 37, 40, 41, 42

Singing 3, 29, 39, 75

Gospel Rap 39

Symbolism 29

Writing 29

Mystery Religion 13, 25, 36, 42, 52, 59

N

Nature 4

Ecological Crises 17, 18, 23, 44, 58, 77

Famines 23

Pandemics 11, 23, 74

Nature-Stewardship 27, 79

Nature-Worship ...... 22, *See* Ancient Ones, Mother Earth, and Ecological Movement

Northern Africa 51

P

Parasocial Interaction 33

Philosophy 13

Scientism . 17, 24, 31, 52, 53, 76, 78

Physicalism 17, 24

Radical Empiricism 17

Phoenicia 57, 58

Politics 13, 17, 21, 34, 56, 74

Civil Resistance 34, 37, 74

Favoritism 25

Terrorism 23, 38, 54, 56, 74, 77

Transnational Advocacy 36

Possession .... *See* Demonization

Propaganda 68

Active Propaganda

Agitation . 12, 24, 25, 38, 53, 54, 56, 69

Horizontal 73

Integration 12, 21, 24, 39, 54, 55, 56, 58, 59, 69

Irrational 73

Pneumatological 76

Political 74

Rational 73

Sociological 74

Vertical 75

Propaganda Device

Evil Suspicion 12

Name-Calling 74

Sub-Propaganda

Conditioned Reflex 12, 68, 72

INDEX

Myth 12, 68, 72

Psychological Security...10, 23, 39, 56, 80, 81

Psychological Warfare 5, 6, 10, 16, 17, 29, 56, 74, 75

R

Religion........ 13, *See* Civil Religion

Atheism 19, 52, 53

Revolution................10, 60, 70, 71

Rome............21, 48, 49, 50, 55, 56

Rotary International.................1, 4

S

Sacred Arts & Sciences.13, 19, 32, 39, 63

Alchemy 19, 43, 61, 70, 72, 76

Pathways of the gods 20, 23, 50, 52, 55, 56

Sacred Law...........................13, 44

Science 13, 17, 76

Computer Science 4

Ecology...31

Engineering 13

Ingenium...........................13, 67

Physics 18, 30, 31

Physiology 30

Pseudoscience 17, 32, 76

Evolutionary Theory ..21, 24, 25

Racial Superiority Theory 12, 21, 24, 25, 36, 52, 53

Psychology 30, 31

Active Imagination 31

Psychophysics 30, 31, 41

Socio-Behavioral Science.........4, 74

Secret Networks.....5, 6, 23, 42, 43

Serpent (The) 7, 54, 60, 62, 63, 77

Serpent-gods 60

Pythōn....................................56

Sexuality

Pansexuality 22, 56, 59, 84

Pederasty........21, 22, 38, 53, 59, 60

Sexual Love...........................56, 79

Shadow Economics 13, 26, 38

Shining Ones...6, 7, 19, 59, 60, 62

Social Engineering 5, 6, 10, 16, 17, 35, 60, 74, 75

Social Governance ...10, 48, 79, 81

Spiritual Heart....9, 12, 14, 44, 47, 71, 79

Conscience.................71, 76, 79, 80

Spiritual Warfare

Counter-Revolutionary Spiritual Warfare........................ 10, 55, 57

Revolutionary Spiritual Warfare .. 6, 10, 40, 44, 47, 76

Spirituality..................................... 4

Animism.. 18, 22, 26, 51, 52, 53, 54

Scriptural Christianity19, 22, 52, 56

Scriptural Judaism 19, 49, 59

The Mysteries .. 5, 6, 19, 22, 41, 44, 47, 53, 56, 72, 75

T

Technique 13, 67, 68, 72

Artificiality 75

Automatism 76

Autonomy 73

Inner-Impetus 76

Monism.. 75

Rationality................................... 75

Secularization 14, 19, 51, 68

Syncretization..... 14, 19, 50, 51, 53, 55, 68

Universalism................................ 73

Technology 4, 13, 68, 79

Cryptocurrencies 16

Culture-Encoding Technologies 21, 25, *See* Civil Religion

Cybernetics 15

Ecosystems................. 15, 16, 33

Eco-Modification Technologies . 16

Electronic-Digital Information Technologies................ 15, 33, 42

Information Warfare Technologies The Internet..... 3, 16, 33, 35, 54

Non-Physical Technologies .... *See* Mathematics

Psycho-Spiritual War-Fighting Technologies... 44, 56, 61, 65, *See* Divination and Enchantment

U

United Nations 5, 6, 35

United States..... 2, 3, 4, 24, 25, 26, 27, 34, 35, 36, 38, 39, 42, 54, 74

W

West (The)..20, 24, 34, 35, 37, 38, 53, 54

Western Asia 21, 51, 59

Western Europe............. 23, 51, 52

World System (The)...4, 6, 10, 13, 15, 41, 47, 49, 62, 65, 71, 76, 79, 81

Army................... 44, 47, 65, 77, 81

INDEX

Elementals 7, 40, 62

False Gospels.......51, 52, 53, 55, 76

gods of the Nations 7, 21, 22, 41, 42, 48, 56, 70

    Apollo 56

    Baal 58, 59

    Osiris...............................62, 63

    Tammuz 59, 60

Mother goddess....6, 58, 60, 62, 63, 72, 77

    Anat 58, 63

    Artemis 56, 63

    Diana 56, 63

    Hathor 63

    Hekat 63

    Isis...63

    Mother Earth. 18, 56, 60, 62, 63

    Netherworld goddess .56, 60, 63

    Queen of Heaven 42, 60, 63

    Sekhmet 63

    Taweret....................................63

Ruling Suprahuman Parties.........61

    god of the World System..7, 49, 63

    Rulers of the World System ...7, 49, 65, 75, 76

The Lie 7, 10, 11, 70, 77, 80, 81

Worldview Warfare 7, 9, 10, 72

    Counter-Revolutionary Worldview Warfare 78, 81, *See* Psychological Security, Social Governance, and Counter-Revolutionary Spiritual Warfare

    Revolutionary Worldview Warfare 11, 13, 14, 15, 20, 21, 22, 23, 25, 31, *See* Psychological Warfare, Social Engineering, and Revolutionary Spiritual Warfare

        Five Fundamental Elements . *See* Revolution, Wicked Deception, Propaganda, Technique, and the Zeitgeist

Z

Zeitgeist 14, 66, 67

    Collective Conscious.14, 15, 24, 36, 53, 67, 76

    Collective Unconscious.. 14, 15, 21, 22, 31, 41, 43, 56, 57, 66, 67, 73, 76

Thank you for reading my book. You can continue following my work at:

www.oliveinformatics.com

Made in the USA
Coppell, TX
07 August 2021